About this book

Framework Maths Year 9S has been written specifically for students who are working below but towards the objectives in Year 9 of the Framework for Teaching Mathematics. The content is based on the Year 8 teaching objectives and leads to the 4–6 tier of entry in the NC tests.

The authors are experienced teachers and maths consultants, who have been incorporating the Framework approaches into their teaching for many years and so are well qualified to help you successfully meet the Framework objectives.

The books are made up of units based on the medium-term plans that complement the Framework document, thus maintaining the required pitch, pace and progression.

This Homework Book is written to support students working below but towards the objectives in Year 9, and is designed to support the use of the Framework Maths 9S Student's Book.

The material is ideal for homework, further work in class and extra practice. It comprises:

- A homework for every lesson, with a focus on problem-solving activities.
- Worked examples as appropriate, so the book is self-contained.
- Past paper SAT questions at the end of each unit, at Level 4 and Level 5 so that you can check students' progress against National Standards.

Problem solving is integrated throughout the material as suggested in the Framework.

Contents

A1.1HW Revising sequences

1 Here are three flow charts.

Follow the instructions to generate a sequence from each flow chart.

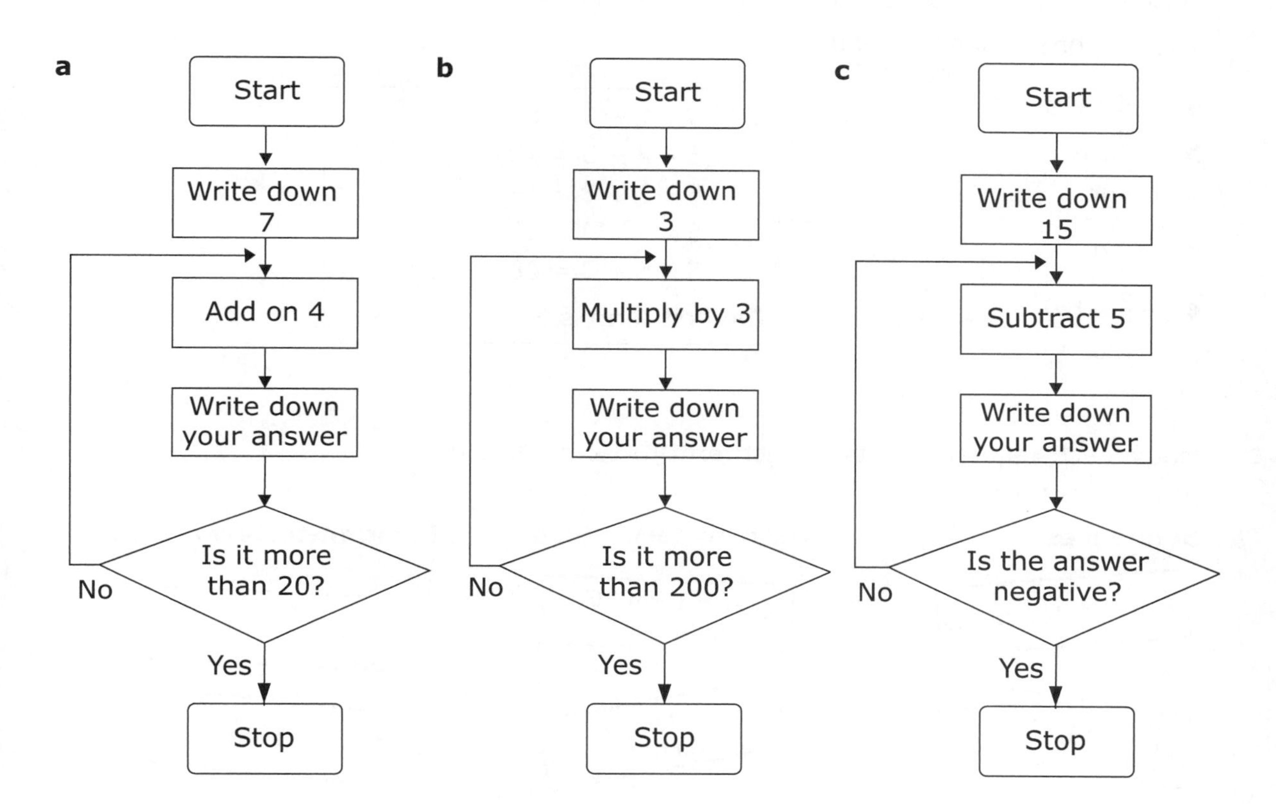

2 Draw a flow chart for generating each of the following sequences:

a 6, 10, 14, 18, 22

b 0.5, 0.75, 1, 1.25, 1.5

c 16, 8, 4, 2, 1, 0.5

A1.2HW Generating sequences

1 The *n*th term of a sequence is given.

Write down the first five terms of each sequence.

Describe each sequence using a term-to-term rule.

The first one is done for you.

a $5n + 3$

b $2n - 1$

c $100 - 10n$

d $\frac{1}{2}n$

e $4n + 0.5$

f $n + \frac{1}{2}$

a $5 \times 1 + 3 = 8$
$5 \times 2 + 3 = 13$
$5 \times 3 + 3 = 18$
$5 \times 4 + 3 = 23$
$5 \times 5 + 3 = 28$

The first term is 8, then add 5.

2 Match each sequence to its correct term-to-term and position-to-term rule.

Sequences	**Term-to-term rules**	**Position-to-term rules**
4, 8, 12, 16, 20, ...	First term 20 Subtract 5	$T(n) = n + 12$
13, 14, 15, 16, 17, ...	First term 3 Add 5	$T(n) = 10n + 15$
3, 8, 13, 18, 23, ...	First term 4 Add 4	$T(n) = 4n$
25, 35, 45, 55, 65, ...	First term 13 Add 1	$T(n) = 25 - 5n$
20, 15, 10, 5, 0, ...	First term 25 Add 10	$T(n) = 5n - 2$

3 Find three different sequences that start with 1, 3, ...

Describe each of the sequences using a term-to-term rule.

A1.3HW Describing sequences

A pond is surrounded by square concrete slabs as shown below:

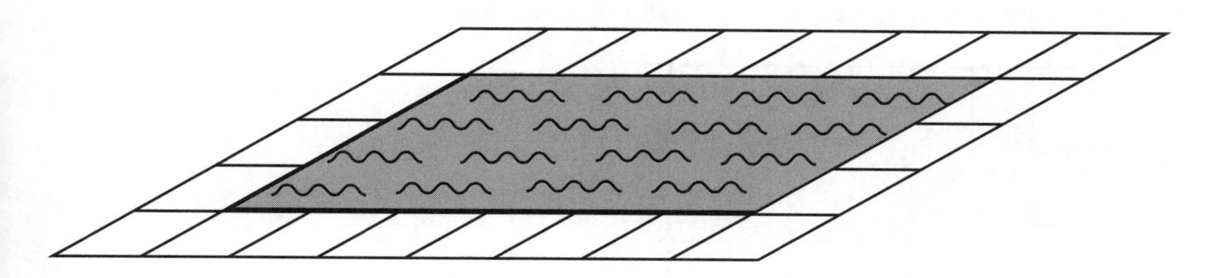

Investigate the number of square concrete slabs surrounding rectangular ponds of different sizes.

For example:

A 2 by 4 pond needs 16 slabs

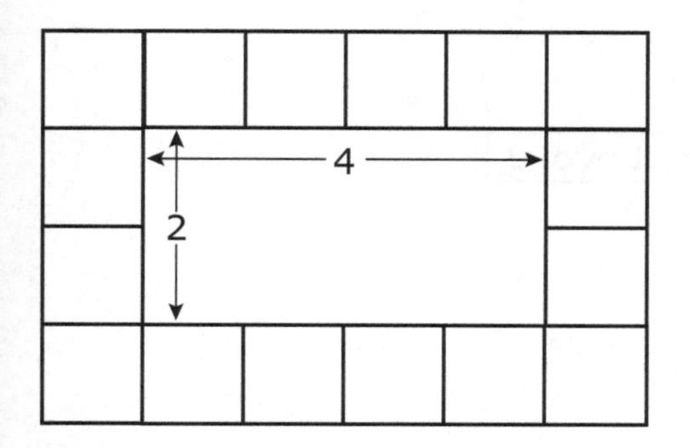

A 1 by 3 pond needs 12 slabs

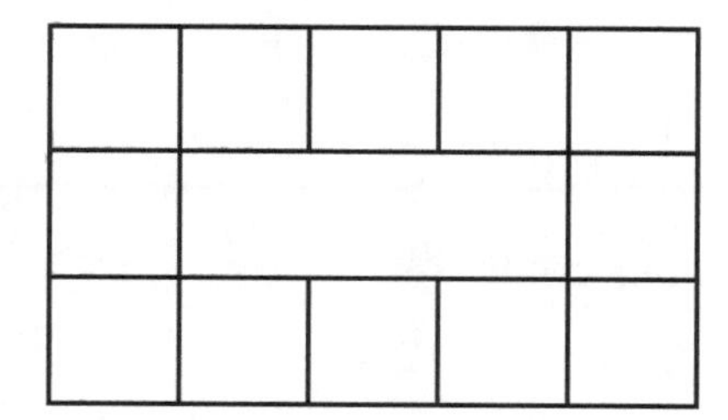

Write a rule, using words or letters, for the number of slabs you will need for any size pond.

Explain why your rule works.

Level 4

Kerry makes a pattern from grey tiles and white tiles.

You cannot see all of the pattern but it continues in the same way.

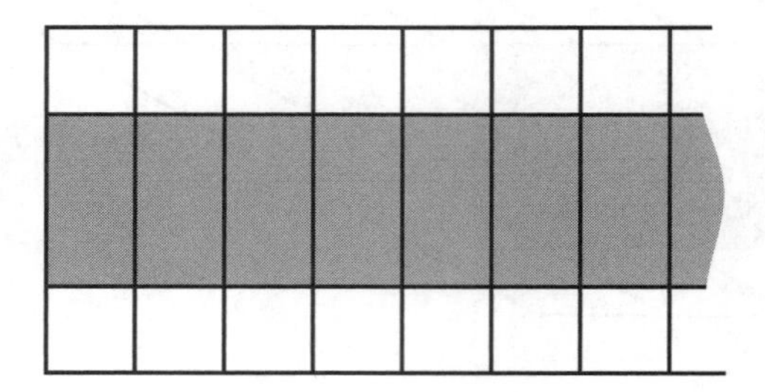

a Kerry uses **30 grey** tiles.

How many white tiles does she use? *1 mark*

b Tim makes a pattern like Kerry's but he uses **64 white** tiles.

How many grey tiles does Tim use? *1 mark*

Level 5

a Here is a number chain:

$2 \rightarrow 4 \rightarrow 6 \rightarrow 8 \rightarrow 10 \rightarrow 12 \rightarrow$

The rule is: **add on 2 each time**.

A different number chain is:

$2 \rightarrow 4 \rightarrow 8 \rightarrow 16 \rightarrow 32 \rightarrow 64 \rightarrow$

What could the rule be? *1 mark*

b Some number chains start like this:

$1 \rightarrow 5 \rightarrow$

Show three **different** ways to continue this number chain.

For each chain write the next three numbers.
Then write the rule you are using. *3 marks*

A2.1HW Revising functions

1 a Complete these function machines and match each one with one of the functions in the boxes. (There are two spare functions.)

i Input x — Output y

1, 2, 3 → $\times 2$ → $+5$ → $^{-}7$, __, __

ii

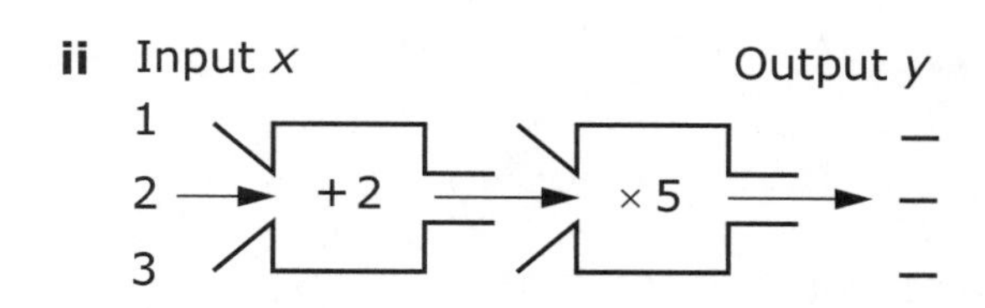

iii

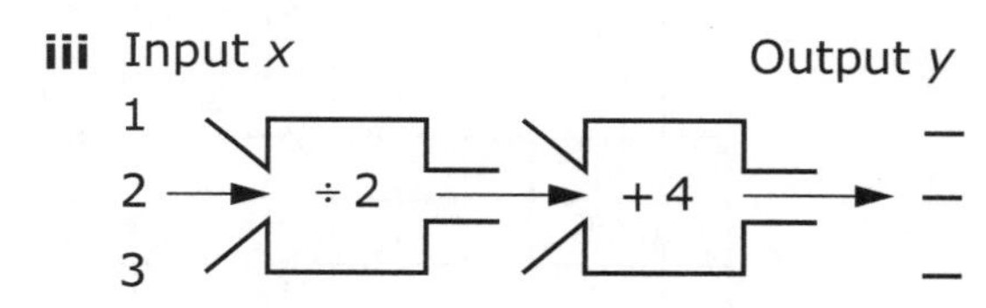

iv

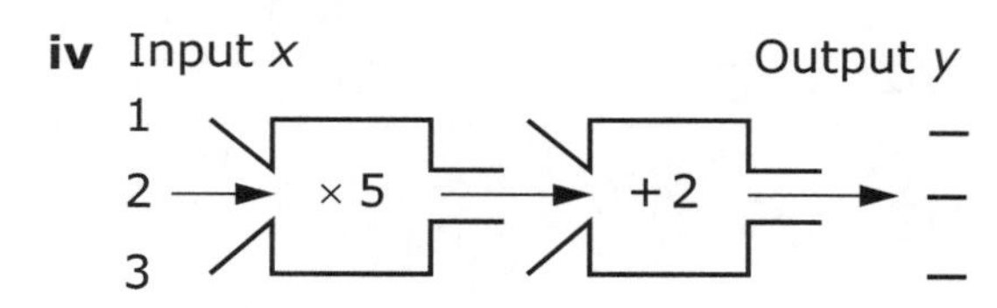

v

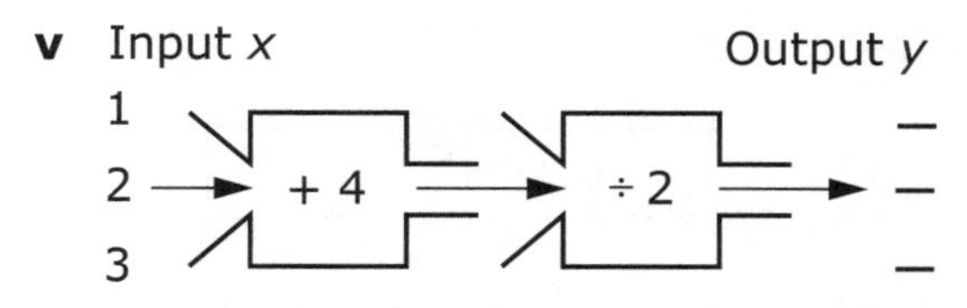

Functions

- $y = \frac{x}{2} + 4$
- $y = 5x + 2$
- $y = \frac{x}{4} + 2$
- $y = 5(x + 2)$
- $y = \frac{x + 4}{2}$
- $y = 2x + 5$
- $y = 2(x + 5)$

b For the two spare functions, draw the function machines.

2 Use any two of these functions … in this function machine:

+4	×2	+2	÷2

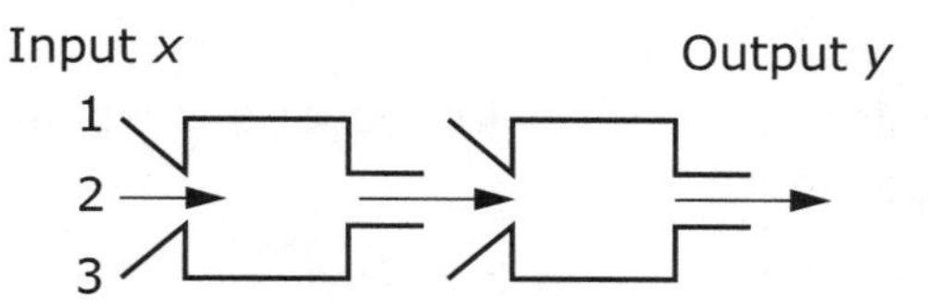

a Which order of functions gives the biggest output?

b Which order of functions gives the smallest output?

c Are there any combinations that give the same outputs?

A2.2HW Drawing graphs from functions

1 For the function $y = 2x + 5$:

a Draw the function machine.

b Use the function machine to generate coordinate pairs.

Input x	Output y	Coordinate pairs
$^-5$	?	(,)
1	?	(,)
3	?	(,)
5	?	(,)

c Copy the grid onto squared paper and plot these points.

Join the points with a straight line.
Label the linear function $y = 2x + 5$.

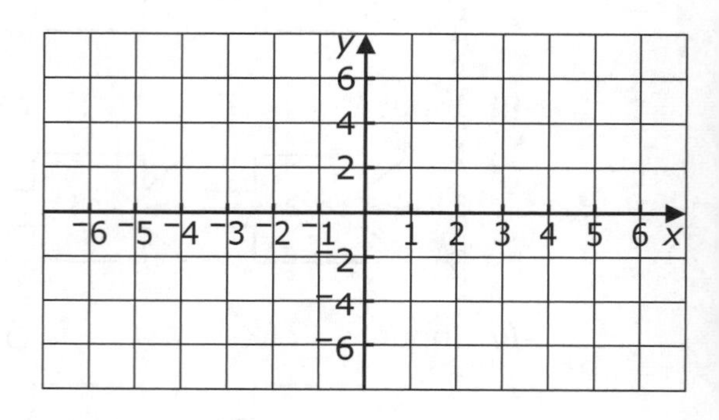

d Use the line to find:

i the output when the input is $x = 2$

ii the input when the output is $y = {}^-2$

2 For the equation $y = 3x - 4$:

a Copy and complete the table of results.

x	$^-2$	$^-1$	0	1	2
y		$^-7$			

b Plot the coordinate pairs on to a grid of suitable size.

c Join up the points and label the line.

3 Which of the following equations would be parallel to the line in question **2**?

$y = 2x + 3$ $\qquad$ $y = 3x + 2$ $\qquad$ $y = {}^-4x + 3$ $\qquad$ $y = 2 + 3x$

A2.3HW Real-life graphs

1 One day in 2004 the exchange rate for £ (sterling) to euros was £1 = 1.6 euros.

Copy the axes for the conversion graph for £ to euros.

Plot the conversion graph using these conversions as a starting point:

£0 = 0 euros

£100 = 160 euros

£200 = 320 euros

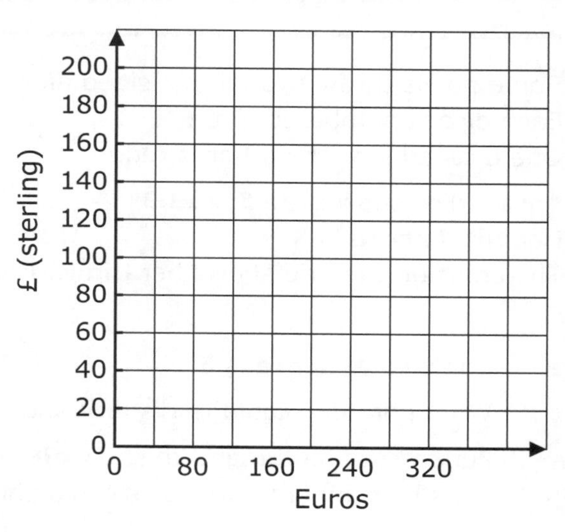

Use your conversion graph to convert the following:

a £5 into euros

b £60 into euros

c £130 into euros

d 240 euros into pounds

e 80 euros into pounds

f 300 euros into pounds.

2 The graph shows the height of the water in the sink as David did the washing up.

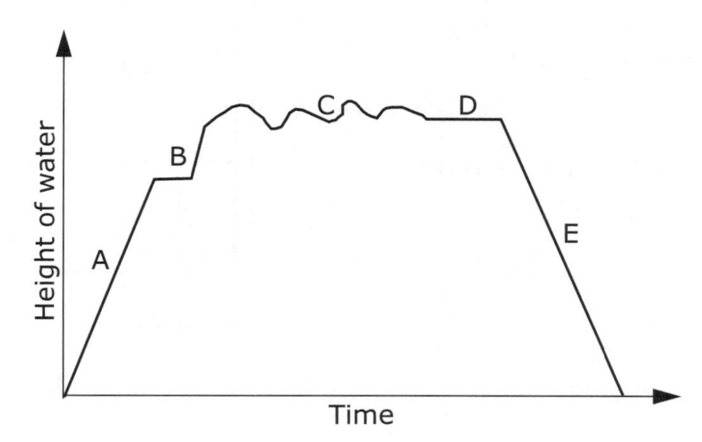

Describe what may be happening at points A, B, C, D and E.

Level 4

Some pupils throw two fair six-sided dice.
Each dice is numbered 1 to 6.
One dice is blue. The other is red.

Anna's dice show **blue 5, red 3**
Her **total score** is **8**
The cross on the grid shows her throw. Copy the grid.

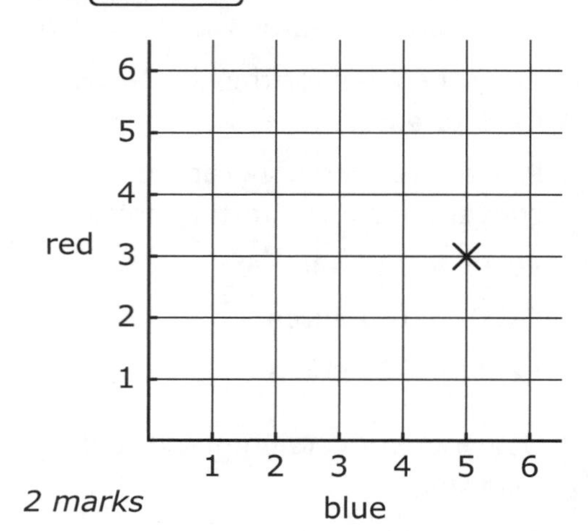

a Carl's **total score** is **6**

What numbers could Carl's dice show?

Put crosses on the grid to show **all** the different pairs of numbers Carl's dice could show. *2 marks*

b The pupils play a game.

Winning rule	Win a point if the number on the **blue** dice is the **same as** the number on the **red** dice.

Put crosses on the grid to show **all** the different winning throws. *2 marks*

c The pupils play a different game.
The grid shows all the different winning throws.

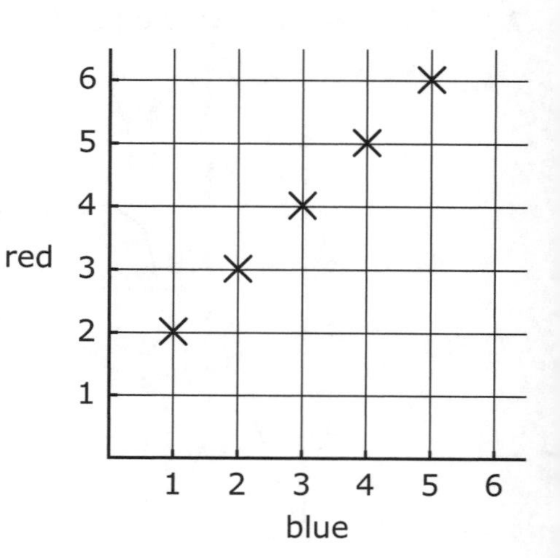

Copy and complete this sentence to show the winning rule.

Winning rule	Win a point if the number on the **blue** dice is ....

1 mark

Level 5

Some people use **yards** to measure length.

The diagram shows one way to change yards to metres.

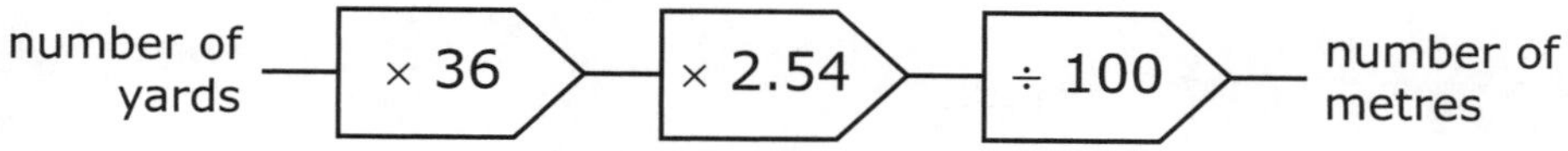

number of yards → $\times 36$ → $\times 2.54$ → $\div 100$ → number of metres

a Change **100 yards** to metres. *1 mark*

b Change **100 metres** to yards.
Show your working. *2 marks*

N1.1HW Adding and subtracting fractions

Remember:

- You can compare fractions when they have the same denominator.

Sammy says that this 5 × 6 rectangle with 30 squares can be split into the following fraction sum:

$$\frac{1}{2}+\frac{1}{5}+\frac{1}{6}+\frac{1}{15}$$

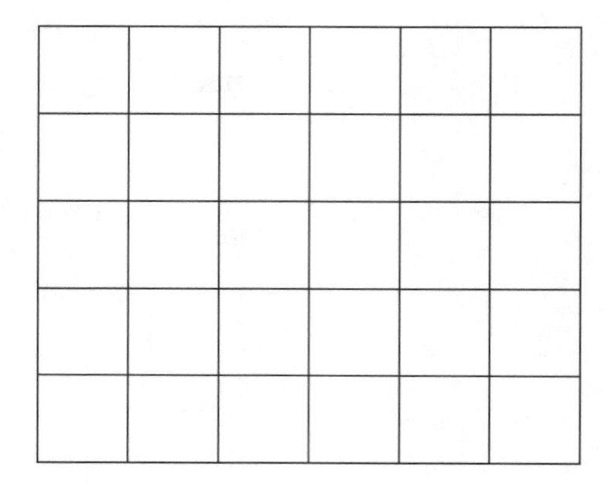

a Copy the 5 × 6 grid and show that this is not true.

b Work out the fraction sum. (Remember that 2, 5, 6 and 15 all divide into 30!)

c Write down what you think the fraction sum should be changed to in order to be correct.

d Find a fraction sum for the 5 × 6 grid that contains

i 2 fractions

ii 3 fractions.

N1.2HW Multiplying by fractions

1 Work out each of the following. The first two are done for you.

a $\frac{1}{3} \times 2 = \frac{2}{3}$

b $3 \times \frac{2}{7} = 3 \times \frac{1}{7} \times 2$

$= \frac{3}{7} \times 2$

$= \frac{6}{7}$

c $2 \times \frac{2}{6} =$

d $\frac{2}{5} \times 2 =$

e $\frac{1}{8} \times 5 =$

f $1 \times \frac{3}{4} =$

g $\frac{1}{6} \times 3 =$

h $2 \times \frac{1}{7} =$

i $\frac{1}{5} \times 2 =$

j $3 \times \frac{2}{8} =$

2 Put the appropriate < (less than) or > (greater than) sign between these statements. (Use the fraction wall to help you.) The first question is done for you.

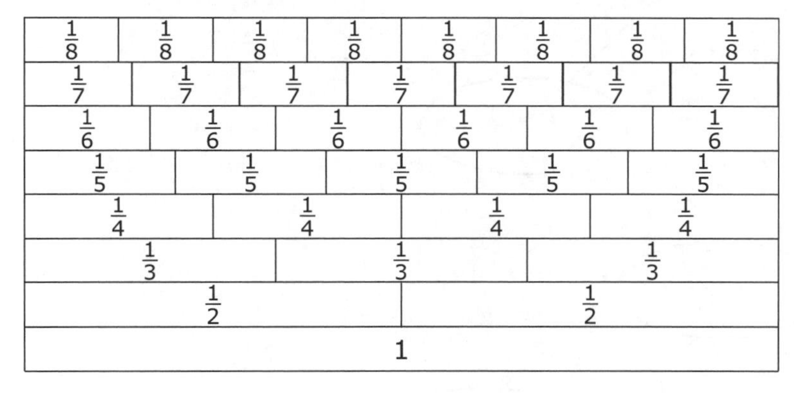

a $3 \times \frac{1}{4}$ __>__ $2 \times \frac{2}{6}$

b $\frac{1}{3} \times 2$ ______ $3 \times \frac{2}{7}$

c $5 \times \frac{1}{8}$ ______ $2 \times \frac{2}{5}$

d $2 \times \frac{1}{5}$ ______ $3 \times \frac{1}{4}$

e $4 \times \frac{1}{8}$ ______ $2 \times \frac{1}{3}$

f $2 \times \frac{1}{6}$ ______ $2 \times \frac{1}{5}$

g $2 \times \frac{3}{8}$ ______ $2 \times \frac{2}{6}$

h $\frac{1}{5} \times 2$ ______ $3 \times \frac{2}{7}$

i $2 \times \frac{1}{5}$ ______ $\frac{1}{6} \times 3$

j $\frac{1}{3} \times 2$ ______ $2 \times \frac{1}{6}$

k $3 \times \frac{1}{4}$ ______ $5 \times \frac{1}{8}$

l $2 \times \frac{1}{7}$ ______ $2 \times \frac{1}{5}$

N1.3HW Dividing by fractions

Remember:

- When you divide an amount by a fraction less than 1, the result will be larger.

Copy and complete these fraction division spider diagrams.
The bubbles without numbers are for your own examples.

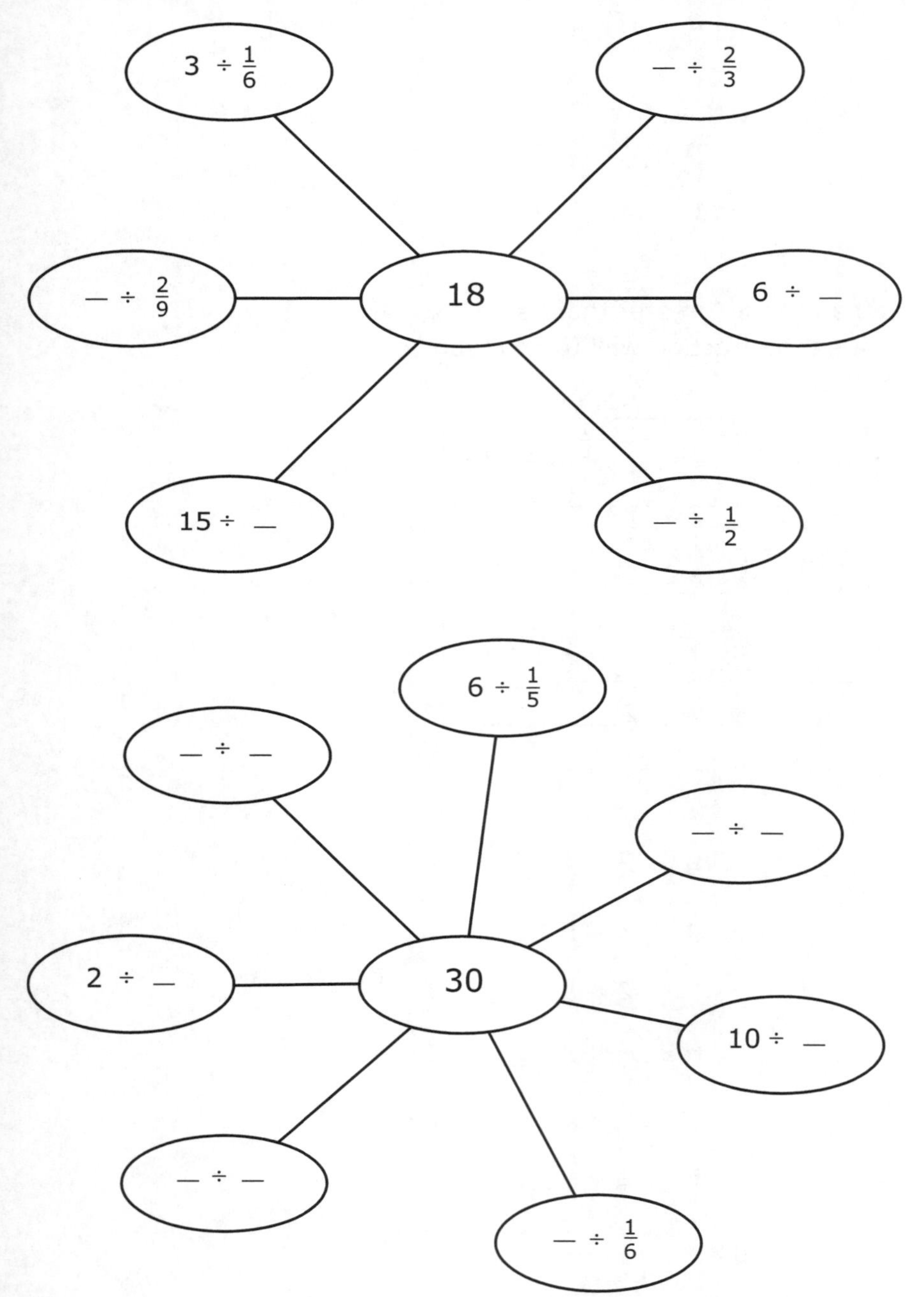

N1.4HW Finding percentages

In these number pyramids, two boxes next to each other add up to make the box on top.

1 Copy and complete this number pyramid using percentage statements.

Instead of writing just the number,
write a percentage statement with the value of that number.

One box is done for you as an example.

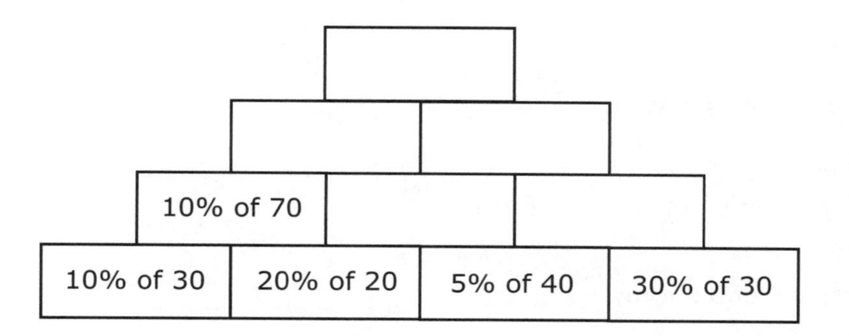

Hint:
10% of 30 = 3
20% of 20 = 4
3 + 4 = 7
7 = 10% of 70

2 Copy and complete this number pyramid.

Write percentage statements so that starting from the bottom, the boxes add up to give the final answer 100 at the top.

Note: There are many answers to this question.

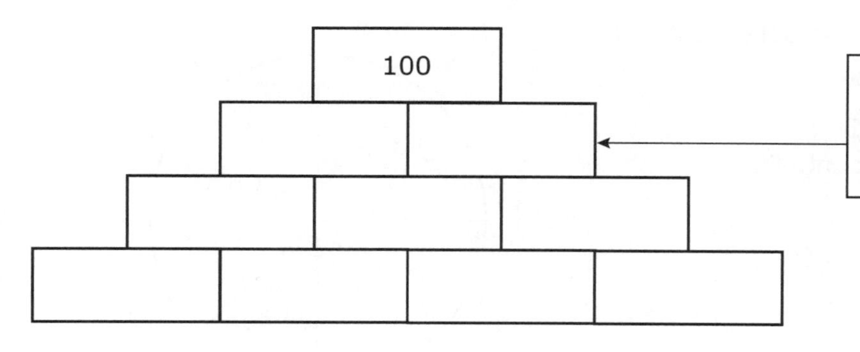

Hint: Start by writing two percentages in these boxes whose answers add to 100.

Fractions, decimals and percentages

1 What fraction of 10 hours is 2 hours?

Express your answer in its simplest form.

2 What percentage of 5 hours is 15 minutes?

3 Jim makes a fruit squash out of fruit juice and lemonade.
If 40% is fruit juice, what fraction is lemonade?
Express your answer in its simplest form.

4 $\frac{1}{20}$ of whole-fat milk is fat.
What percentage of whole-fat milk is **not** fat?

5 A shop is offering 15% off in a sale.
Express this discount as a fraction in its simplest form.

6 This bar chart shows the results of a survey about how students travel to school.

a What percentage travel by bus?

b What fraction walk or come by car?
(Express your answer in its simplest form.)

c What fraction use wheels as a method of transport?
(Express your answer in its simplest form.)

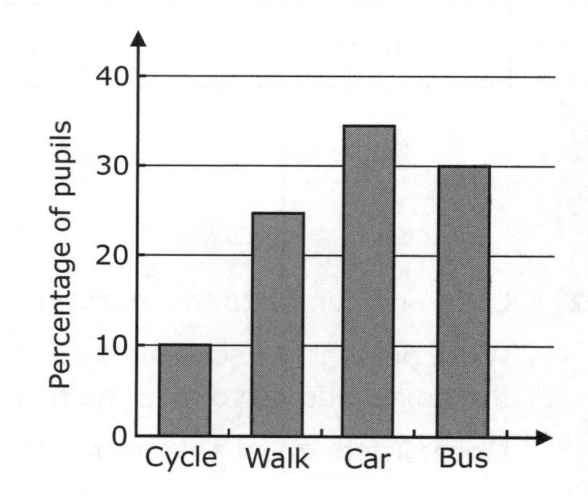

7 This pie chart shows students' favourite fizzy drink flavours.

80 students liked orange.

Estimate how many students liked:

a Cola

b Lemon.

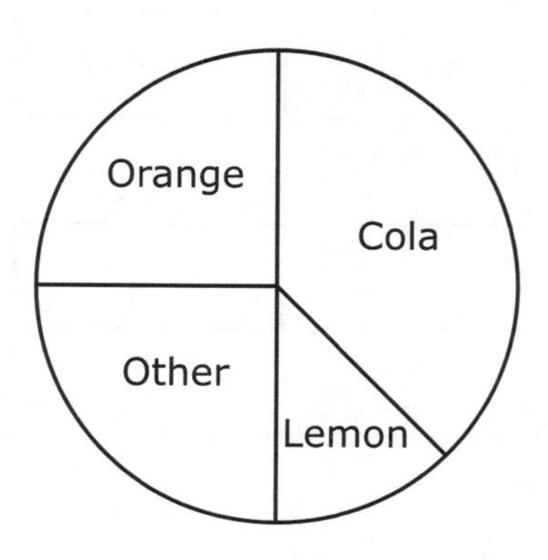

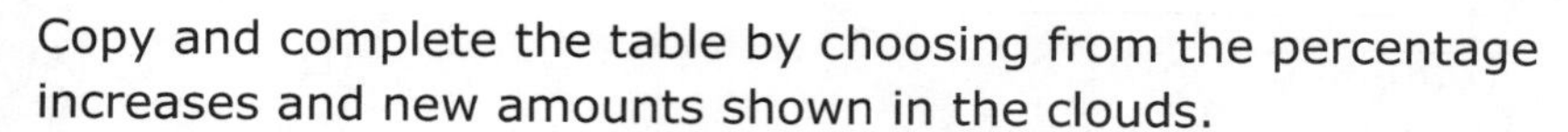

N1.6HW Percentage change

Copy and complete the table by choosing from the percentage increases and new amounts shown in the clouds.

The first one has been done for you.

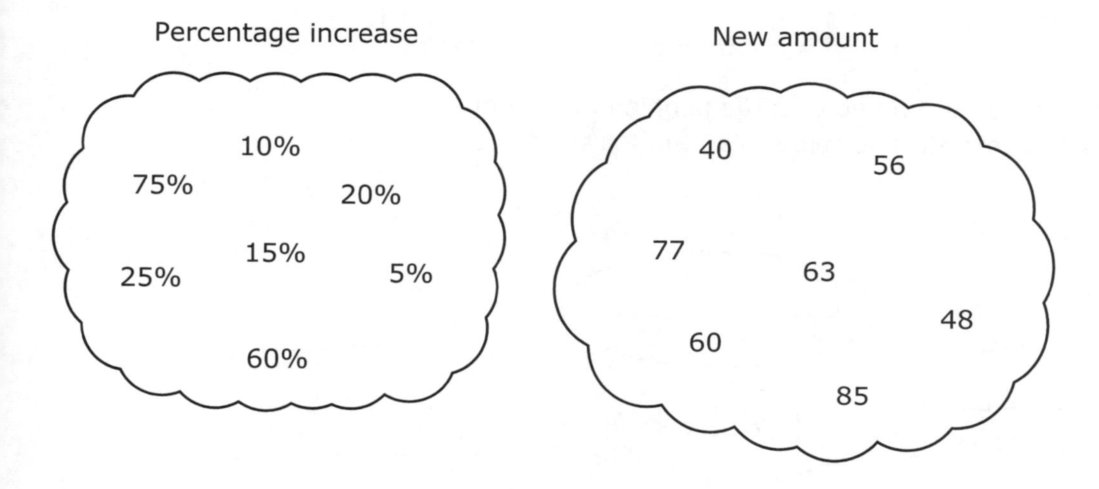

Amount	% increase	New amount
40	20%	48
32		
70		
68		
25		

Remember:

To find the new amount, add the increase on to the original amount.

N1.7HW Using ratio

Remember:

- You can simplify a ratio by dividing each part by the same number.

  For example 4:2 = 2:1

1 £72 can be shared between two or three people in different ratios. Copy and complete this spider diagram for sharing £72 in each ratio.

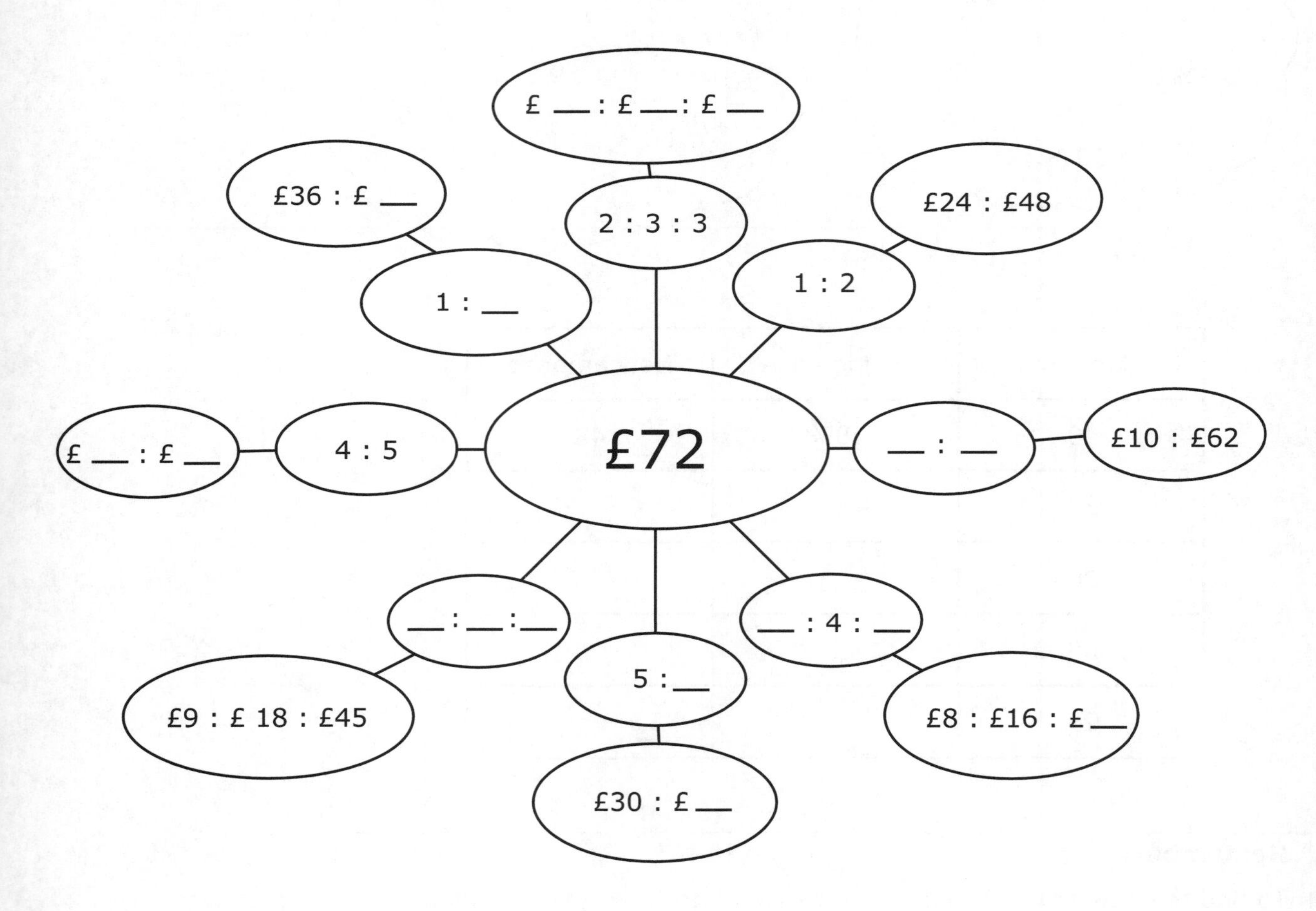

2 Make up your own spider diagram for £40.

N1.8HW Direct proportion

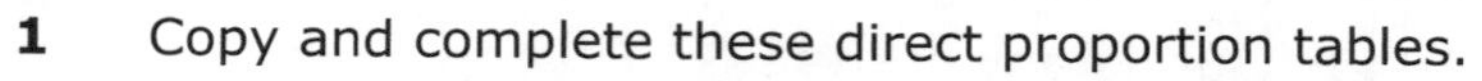

1 Copy and complete these direct proportion tables.

a 1 gallon = 8 pints

Gallons	1	2	5	10	20	50
Pints	8		40			

b 1 litre = 1.76 pints

Litres	1	2	3	7	11	25
Pints		3.52			19.36	

c 1 kilogram = 2.2 pounds

Kilograms	1		21		77	
Pounds		22		72.6		264

2 Use a conversion of your own to complete this table.

1 ___________ = ___________

___________	1	2	4	10	25	90
___________						

3 A 125 g jar of coffee costs £2.75.

A 240 g jar of the same coffee costs £4.95.

a Work out the cost of 100 g of coffee for:

i the 125 g jar

ii the 240 g jar.

b Which jar represents the best value?

Explain your answer using your calculations for part **a**.

> **Hint:** Use the unitary method – find the cost of 1 g.

N1.9HW Order of operations

1 Combine the four digits 2 4 6 8 using any of the operations to make these totals.

Write out your methods using brackets where you need to.

a 100 **b** 28 **c** 4 **d** 20 **e** 8

2 Work out the four calculations in each grid and find which calculation is the odd one out.

a

$10 + 30 \div 3$	$25 - 40 \div 8$
$5 + 4 \times 4 + 4$	$80 \div (8 \div 2)$

b

$(2 + 5) \times (3 - 1)$	$3 \times (4 + 5) - 15$
$10 + 3 \times 2 - 4$	$4 \times (2^2 - 1)$

c

$3 \times 2 \times 5 \div 2$	$6 \times (3 + 2)$
$6 \times 3 - 3$	$3 \times 4 + 3$

d

$2^2 + 3^2 \times 4$	$100 - 20 \div 2$
$(2 \times 4)\ (7 - 2)$	$2 + (6 \times 5 + (4 \times 2))$

3 Make the number 100, using each of the digits 1 to 9 just once, all in the correct order.

Here is one method:

$$123 - 4 - 5 - 6 - 7 + 8 - 9 = 100$$

Remember to insert brackets if you are using × or ÷ as well as + or −.

Level 4

Some towns and villages have very long names.

The table shows information about the ten longest place names in the UK.

Number of letters	Country
67	Wales
58	Wales
27	England
22	Wales
21	Wales
21	Wales
19	England
18	England
18	Scotland
17	Scotland

a The longest place name in **Wales** has more letters than the longest place name in **Scotland**.

How many more? *1 mark*

b **50%** of the ten longest place names are in Wales.

What percentage of the ten longest place names are in **England**? *1 mark*

Level 5

You can make different colours of paint by mixing red, blue and yellow in different **proportions**.

For example, you can make green by mixing **1 part blue** to **1 part yellow**.

a To make purple, you mix 3 parts red to 7 parts blue.

How much of each colour do you need to make **20 litres** of purple paint?

Give your answer in litres. *2 marks*

b To make orange, you mix **13 parts yellow** to **7 parts red**.

How much of each colour do you need to make **10 litres** of orange paint?

Give your answer in litres. *2 marks*

Design a set of dominoes to help practise simplifying expressions.

For example:

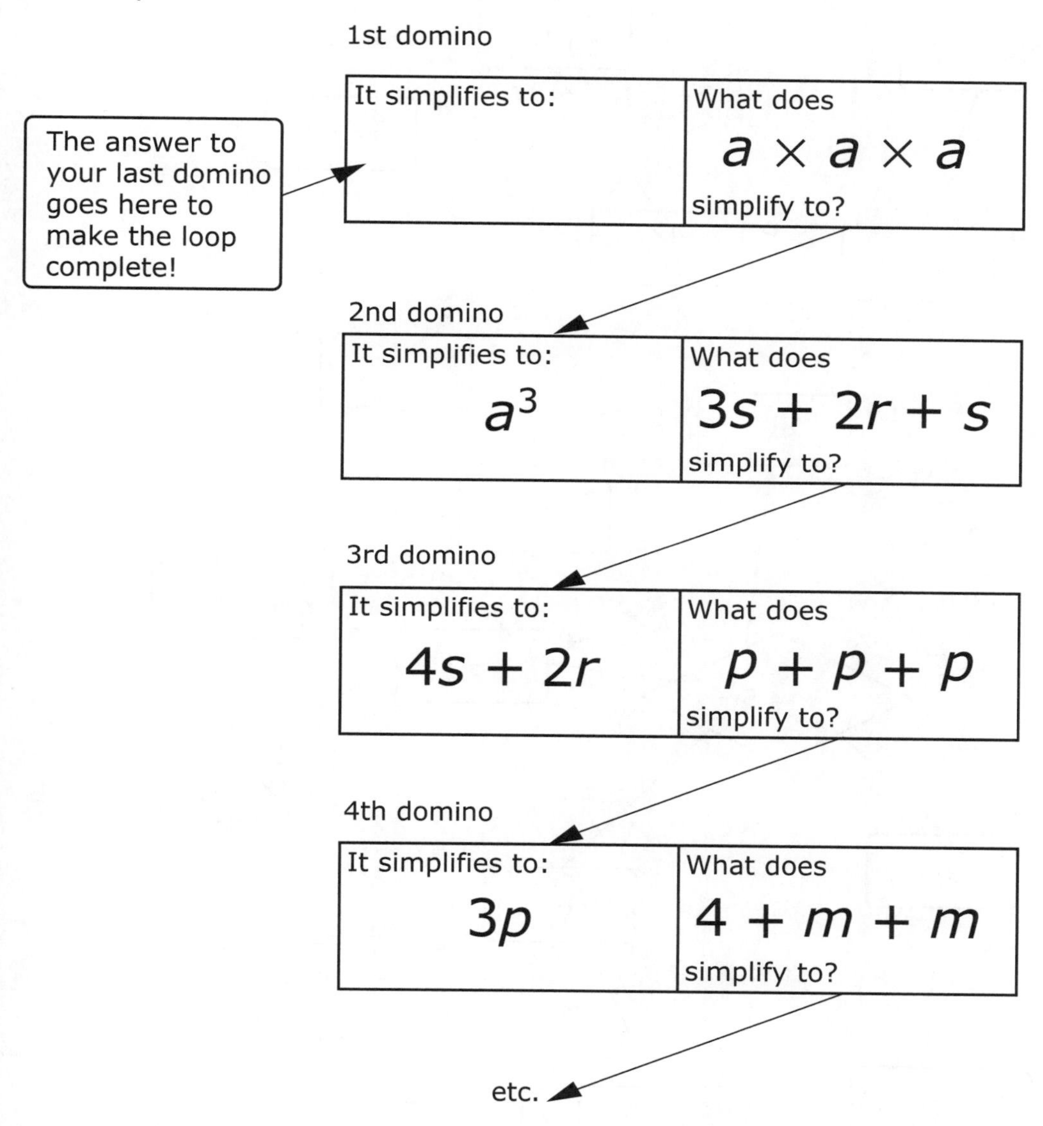

Continue these dominoes or make up your own.

Play a game of dominoes with a partner.

A3.2HW Arithmetic in algebra

Find the value of each expression in this spider diagram when $x = 5$.

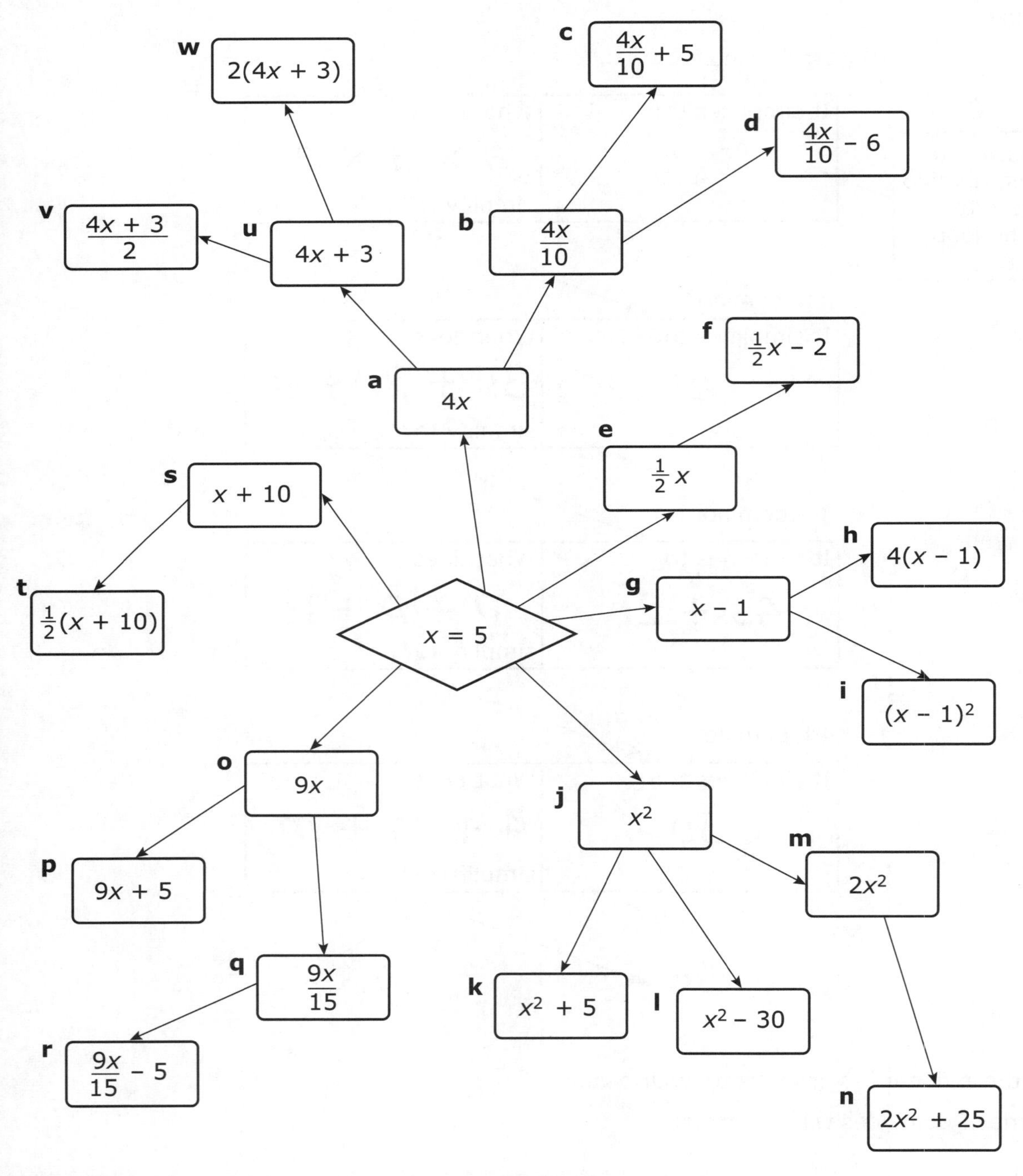

Make up your own spider diagram for someone else to complete.
Make sure you have the answers worked out as well!

Copy and complete the crossword with the solutions to the equations below.
Some of the clues have been done for you.

a 4	■	*b* 1	■	■	*d*	*e*
■	*c* 2	5	■	*f*	■	
g	■	■	*h*		*j*	■
i	*m*	*n*	■	■	*k*	*q*
■	*p*		*s*		■	
r	■	■		■	*v*	■
t			■	*u*		■

Across

$a + 6 = 10$

$20 + c = 45$

$d + 45 = 100$

$\frac{h}{2} + 10 = 91$

$i + 50 = 295$

$\frac{k}{2} = 12$

$2p = 3500$

$2u + 4 = 70$

$2t + 200 = 1000$

Down

$2b + 4 = 34$

$2e + 4 = 110$

$\frac{f + 4}{2} = 35$

$3(g - 2) = 90$

$4j + 2 = 90$

$2m - 2 = 80$

$2(n + 3) = 120$

$2q = 80$

$\frac{r + 4}{2} = 14$

$3s - 50 = 100$

$3v - 9 = 30$

A3.4HW Balancing linear equations

In a two-way flow diagram, you reach the same finish number F, whichever route you follow.

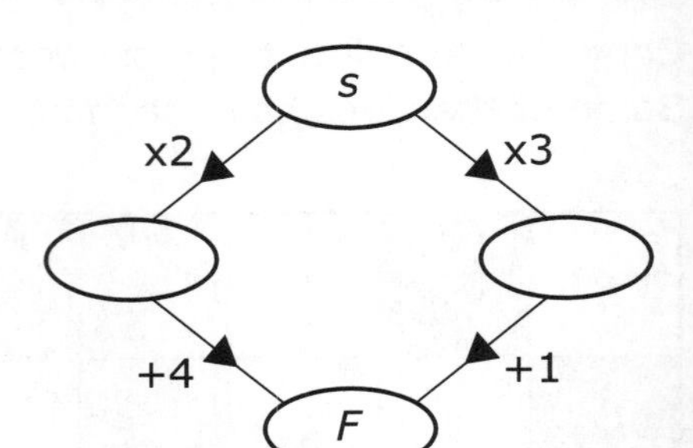

For example:

Using the left path

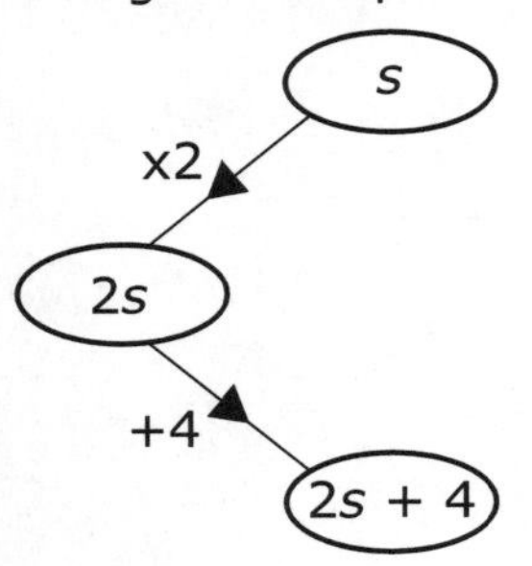

Using the right path

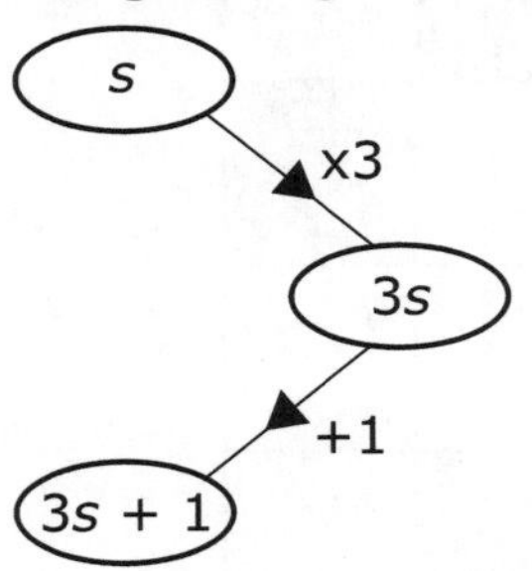

So, $F = 2s + 4 = 3s + 1$

Take off $2s$ from both sides $\quad -2s \quad -2s$

$4 = s + 1$

Take off 1 from both sides $\quad -1 \quad -1$

$3 = s$

So, $s = 3$ and $F = 3s + 1 = 10$.

Find s and F in these two-way flow diagrams by forming equations.

a

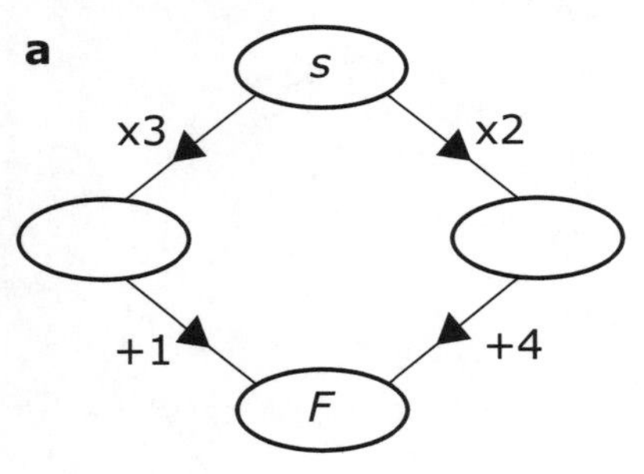

b

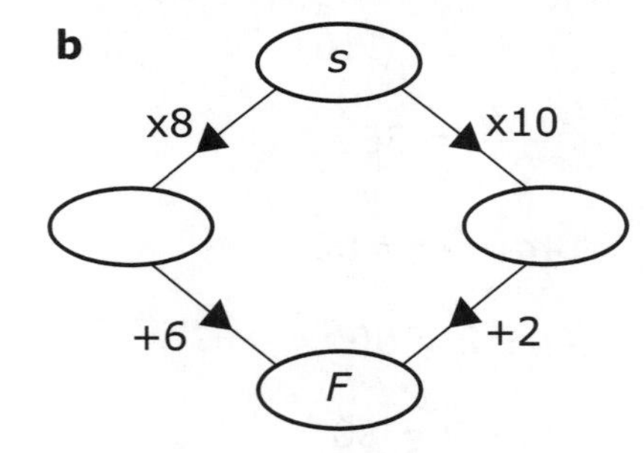

c

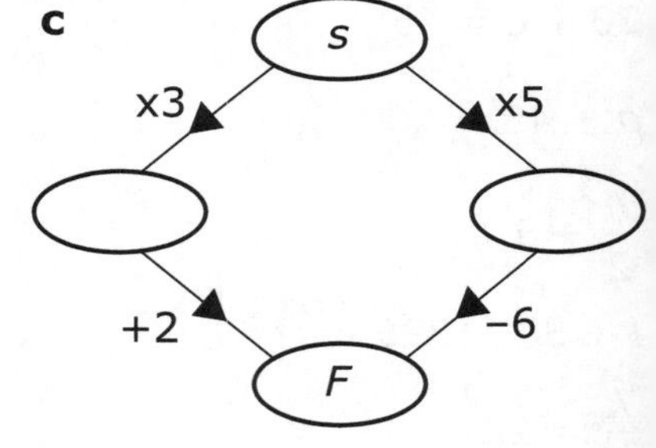

d

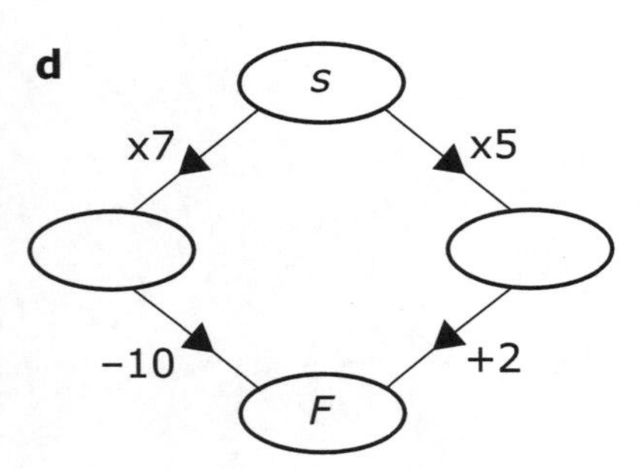

e

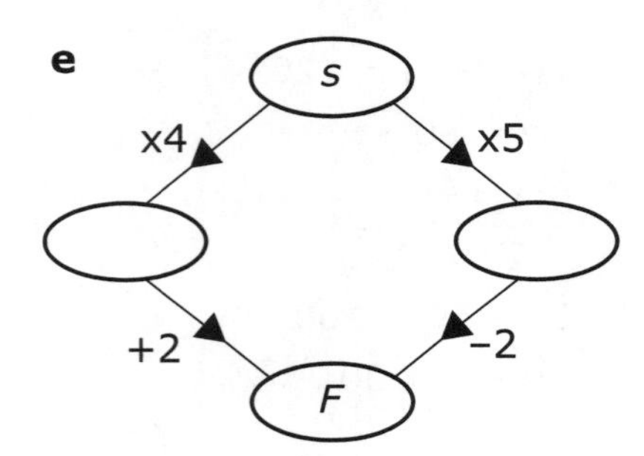

f

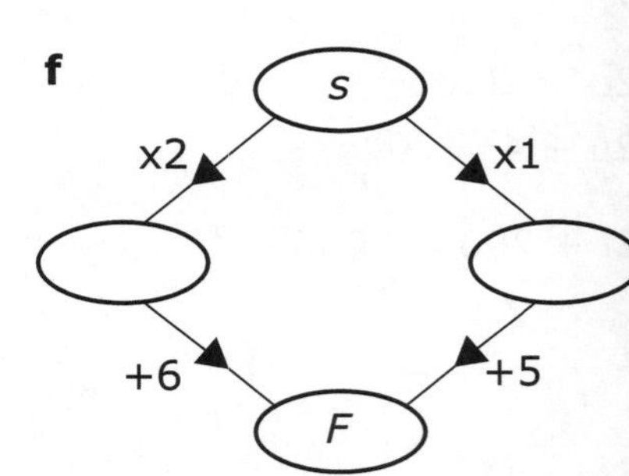

A3.5HW Solving proportional problems

These pictures below are all enlargements, or reductions, of the first picture.
Some are not in the same proportion.
Work out which pictures are not in the same proportion by drawing a graph on axes like this.

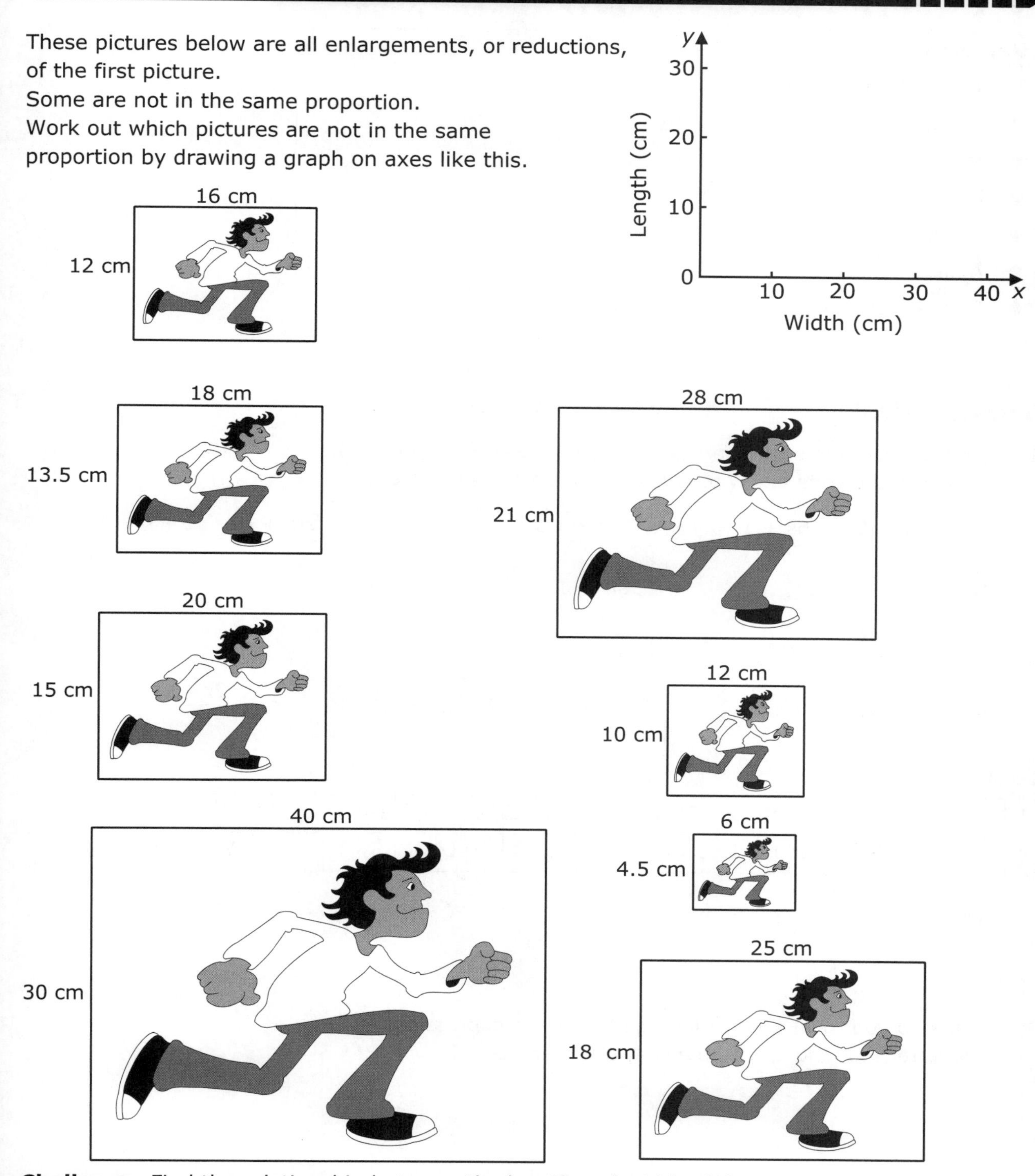

Challenge: Find the relationship between the length and width of the photographs that are in the same proportion, and write this as a formula.

A3.6HW Substitution

Remember:

The order of operations is:

- Brackets
- Indices
- Division
- Multiplication
- Addition
- Subtraction

Secret codes!

1 Find the value of each expression when $r = 3$, $s = 2$ and $t = 4$.
Then use the code below to change your answers to letters.

a $2t$ $5r$ $6s + 1$ $r + s$ $t(r + s) + r$ $3(r + s)$ $r(s + t)$ $t + r + 2s$

b $\frac{t^2}{2}$ $t + 1$ s^2r t^2 $sr^2 + 1$

c $r^3 - s$ $r(t + 1)$ $5t + 1$

d $(s + 1)^2$ $r^2 + t$ s^4 $rt + rs$ $2r + 2t + 1$ $t^2 + 2r$ $\frac{2t + s}{2}$

Code

1	2	3	4	5	6	7	8	9	10	11	12	13	14	15	16	17	18	19	20	21	22	23	24	25	26
A	B	C	D	E	F	G	H	I	J	K	L	M	N	O	P	Q	R	S	T	U	V	W	X	Y	Z

What does the message say?

2 Make up your own secret message by writing expressions with the letter a, b and c, where $a = 4$, $b = 2$ and $c = 5$.

Level 4

a I think of a number. I call my number n.

n

Then I add 5 to my number.

$n + 5$

The answer is 8.

$n + 5 = 8$

What was my number?

1 mark

b Solve this equation to find the value of m.

$m - 2 = 8$

1 mark

Level 5

A teacher has **5 full packets** of mints and **6 single** mints.

The number of mints inside each packet is the same.

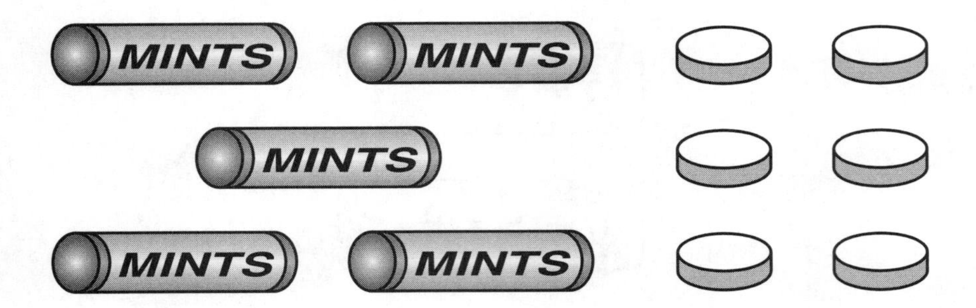

The teacher tells the class:

> '**Write an expression** to show **how many mints** there are **altogether**. Call the number of mints inside each packet *y*.'

Here are some of the expressions that the pupils write:

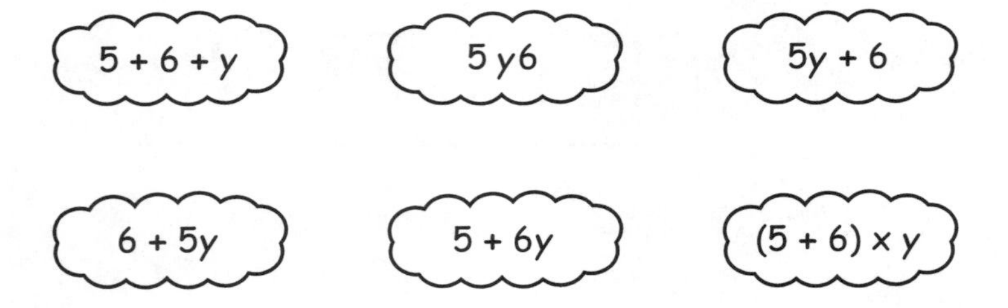

a Write down **two** expressions that are correct. *2 marks*

b A pupil says: 'I think the teacher has a total of **56 mints**'.

Could the pupil be correct? Explain how you know. *1 mark*

S1.1HW Geometrical language

1 Copy and complete each of these sentences.

Choose from this list:
equal
regular
parallel
isosceles
perpendicular

a _______ lines intersect at right angles.

b _______ lines are always equidistant.

c A parallelogram is made up of two pairs of _______ lines.

d A _______ shape has equal sides and equal angles.

e A rhombus can be made using two _______ triangles.

2 Find the value of the unknown angle in these diagrams.

a

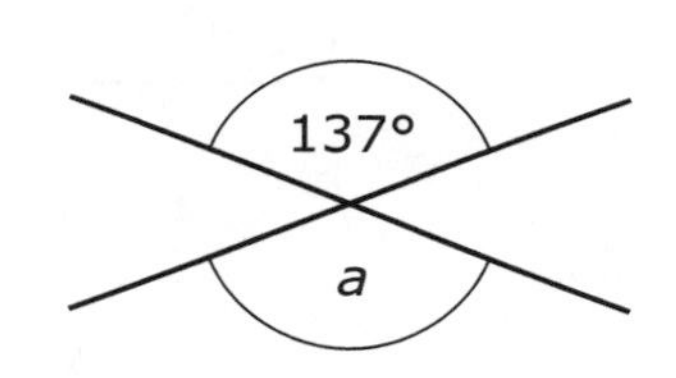

b

3 Rearrange the letters to give a mathematical term.

a LLALAPER

b GONOPLY

c TROUNCENG

d RALUGER

e PONNAGET

4 Draw and cut out two congruent right-angled triangles:

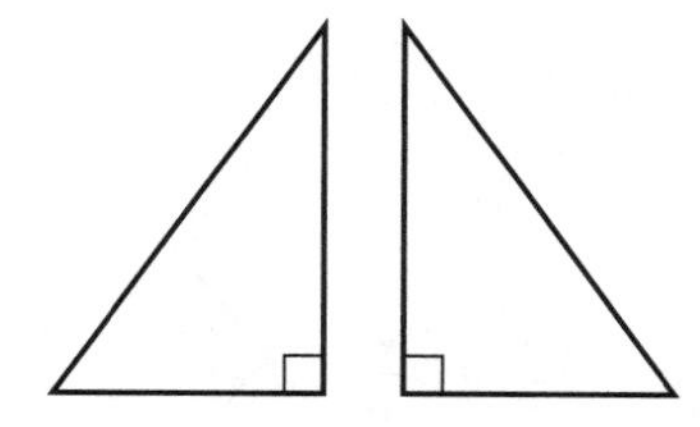

Draw diagrams to show all the shapes you can make by joining the triangles together. Name each shape you make.

S1.2HW Properties of triangles

1 Draw these triangles on an isometric grid.

a equilateral **b** isosceles **c** scalene **d** right-angled

Mark the equal angles and equal sides on your diagrams.

2 Find the unknown angles.

a

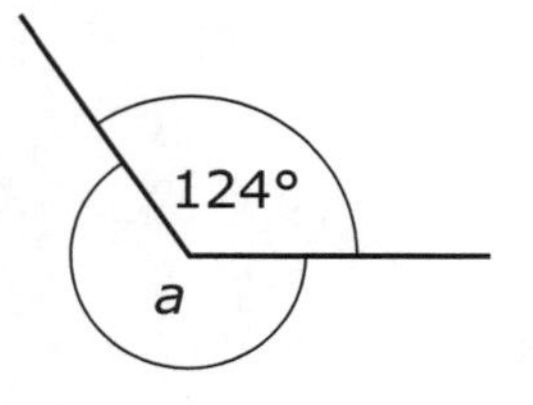

b

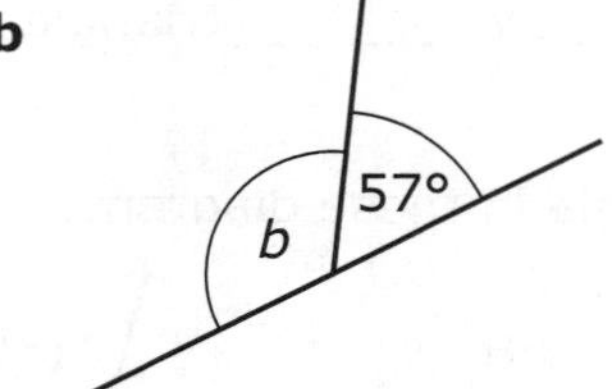

c

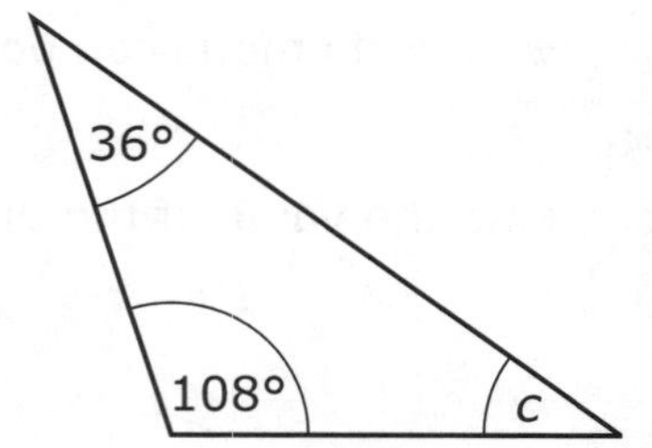

d

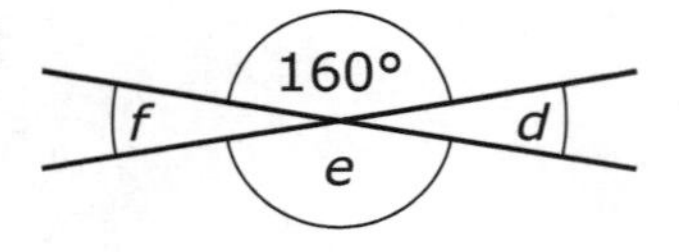

e

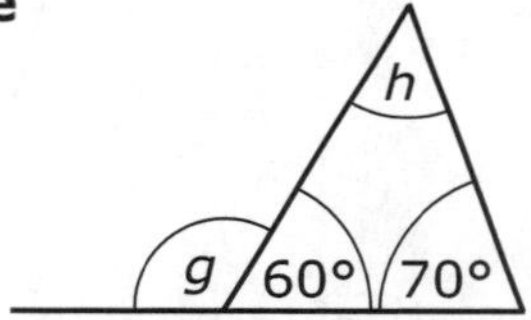

f

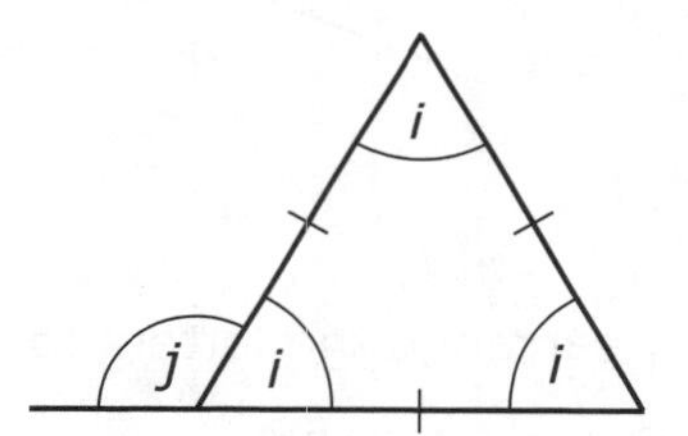

g

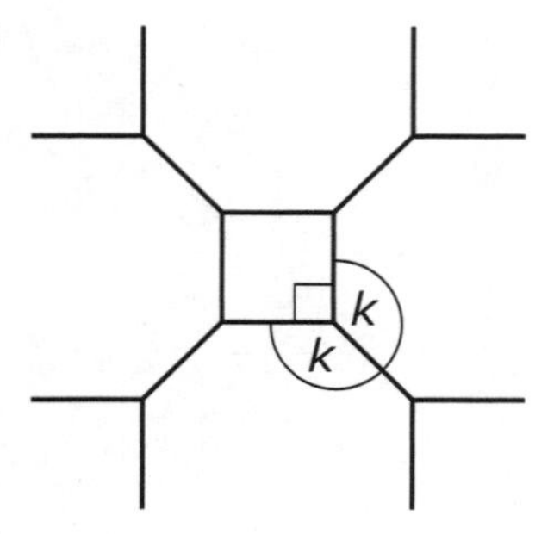

h

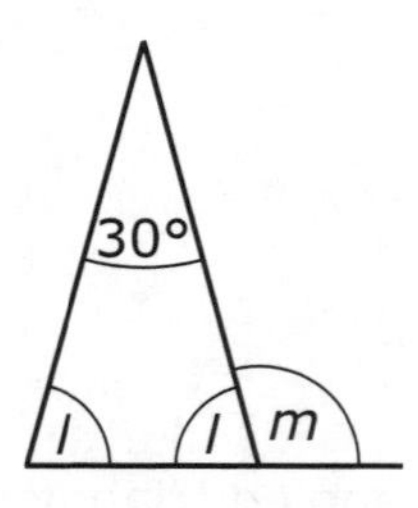

i

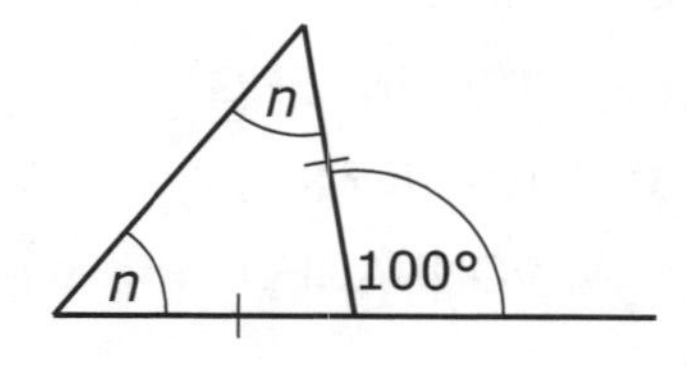

j

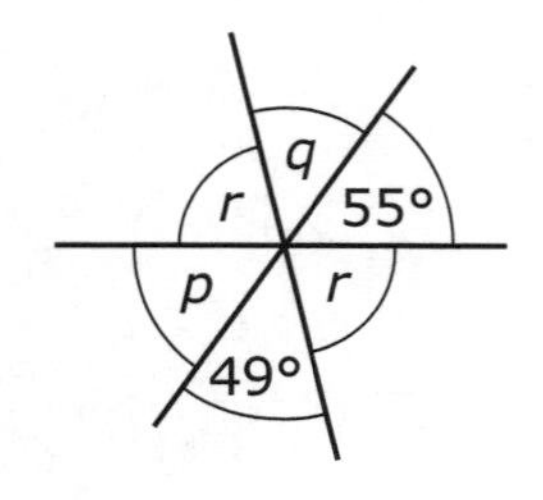

k

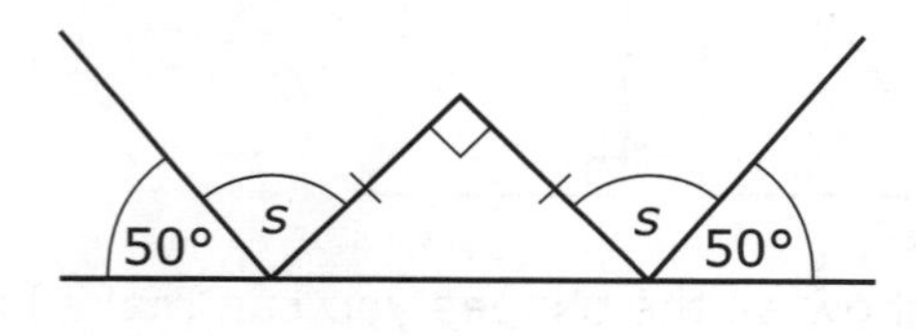

S1.3HW Lines and angles

1 Copy this diagram which shows three sets of three parallel lines.

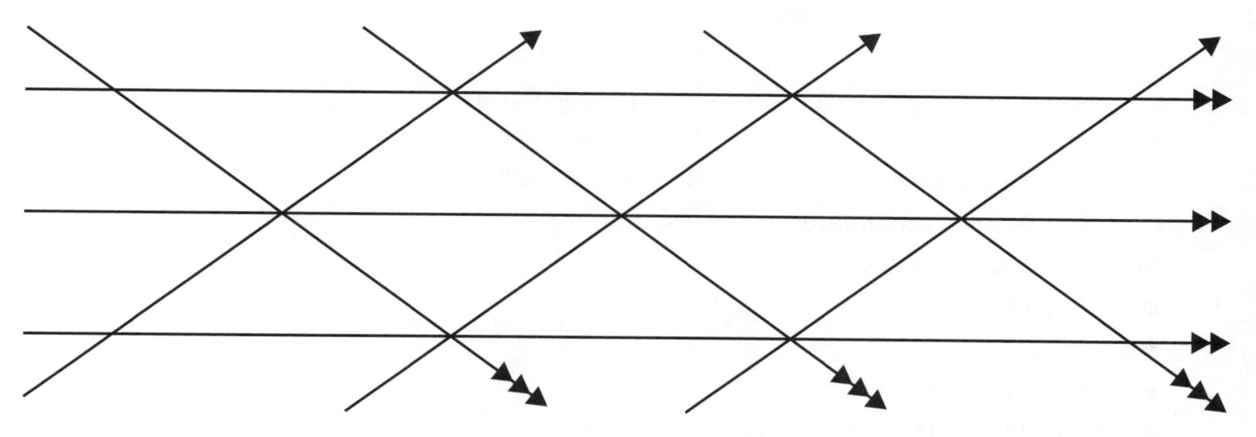

Using three different colours, colour angles that are equal in size.

2
a Which angle is vertically opposite *a*?
b Which angle is alternate to *b*?
c Which angle is alternate to *c*?
d Which angle is corresponding to *e*?
e Which angle is corresponding to *f*?

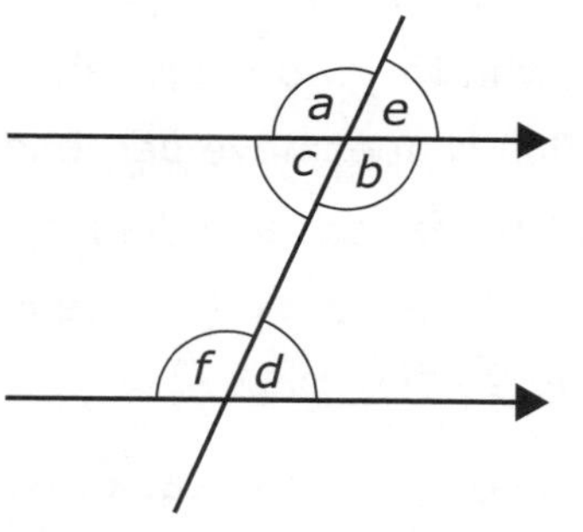

3 Find the unknown angles. Give a reason in each case.

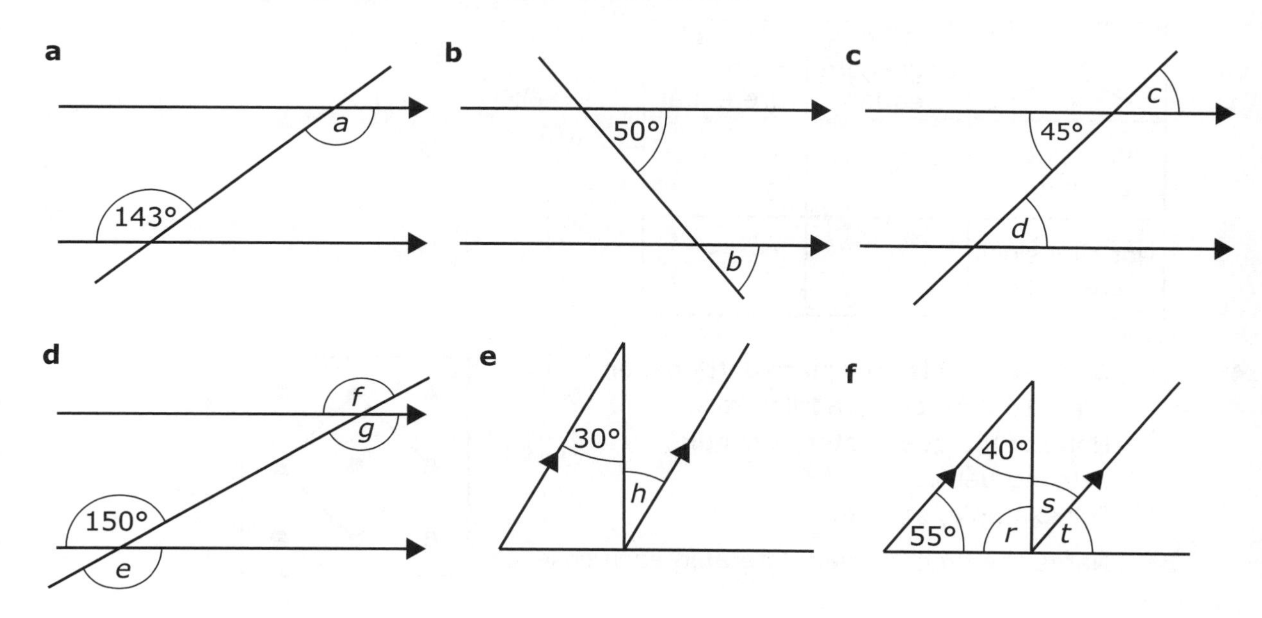

S1.4HW Classifying quadrilaterals

1 Draw these quadrilaterals on isometric paper.
One is impossible – which one?

a square
b rectangle
c rhombus
d parallelogram
e kite
f trapezium
g isosceles trapezium
h arrowhead
i an irregular quadrilateral

On each diagram mark:

- equal angles
- equal sides
- parallel lines
- perpendicular lines (∟).

2 **a** Draw the shapes **a–h** from queston **1** on plain paper.

b Draw in the two diagonals for each shape

c Which shapes have perpendicular diagonals?

d Which shapes have diagonals that bisect each other?

e Which shapes have perpendicular diagonals that bisect each other?

3 Copy and complete this table, writing quadrilaterals from question **2** in the correct place.

	All 4 angles are equal	Not all 4 angles are equal
All 4 sides are equal		
Not all 4 sides are equal		

4 **a** On 3 by 3 grids on square dotty paper, draw 16 different quadrilaterals.
None of the quadrilaterals should be congruent.
For example, square:

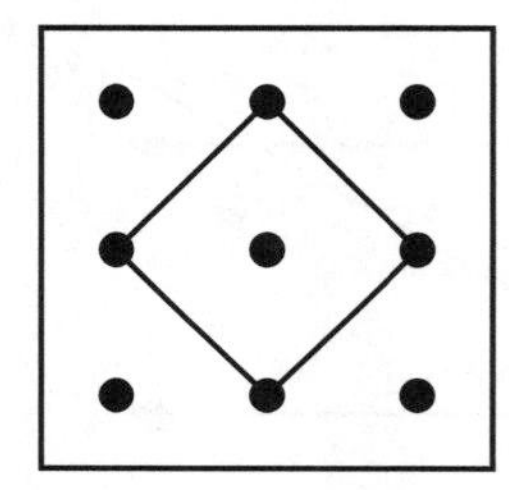

b Name the type of quadrilateral in each case.

Properties of quadrilaterals

Remember: The sum of angles in a quadrilateral is 360°.

1 Find the unknown angles. Give a reason in each case.

a

b

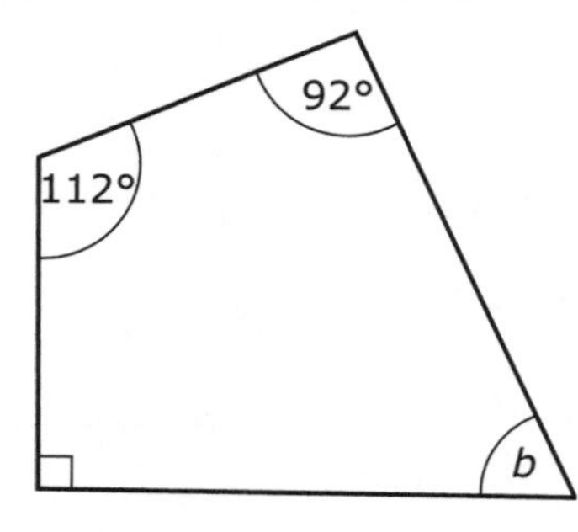

c

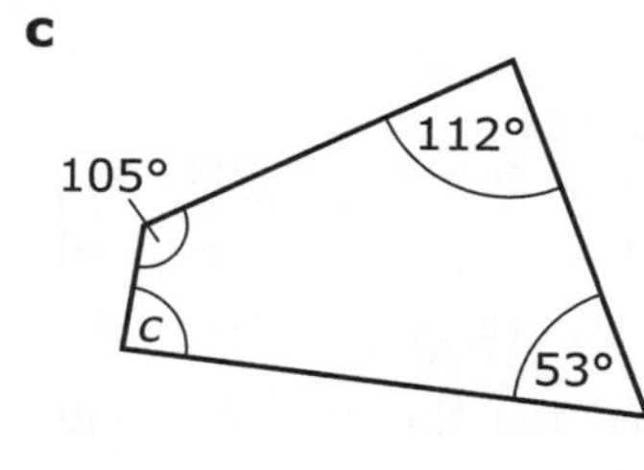

d

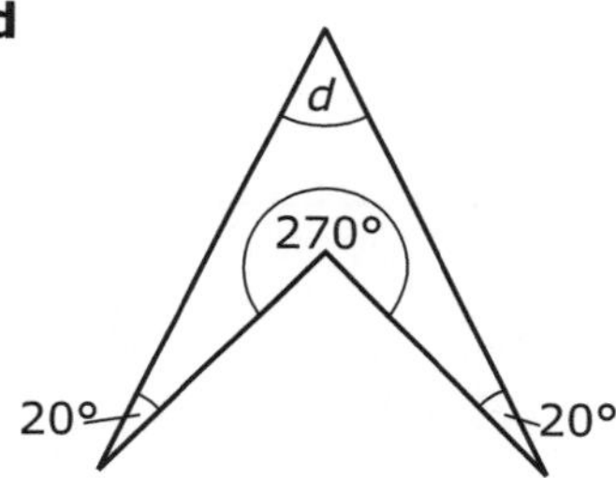

e

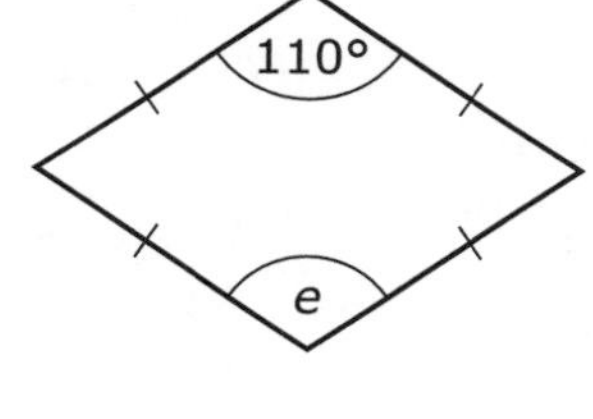

f

2 Find the unknown angles in these tessellations.

a

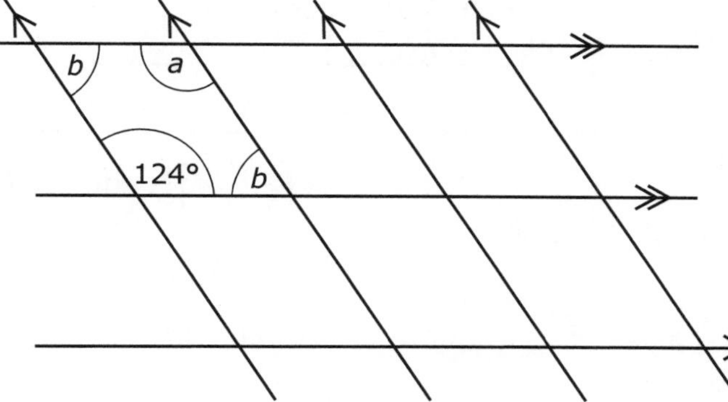

b

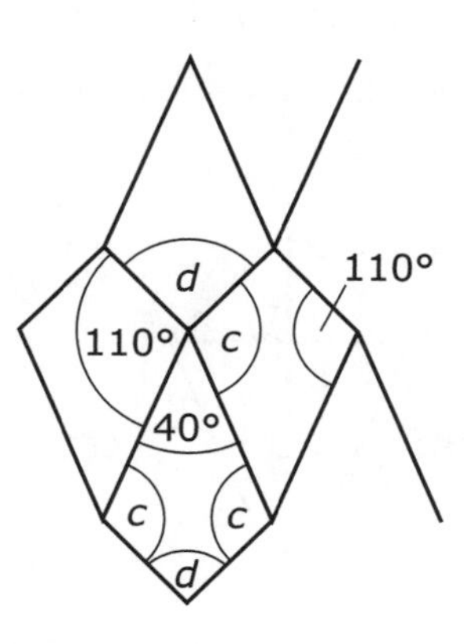

c

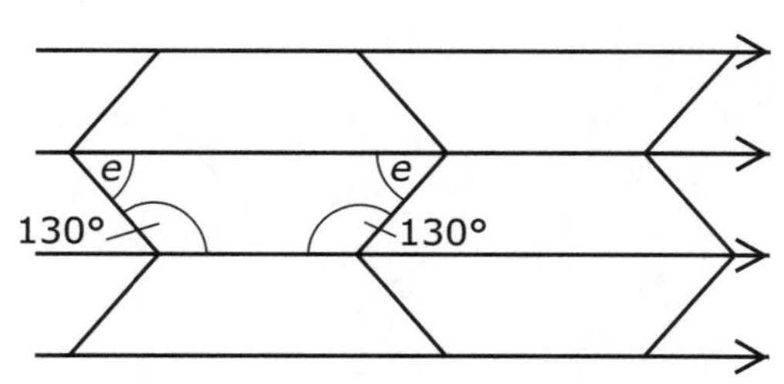

d

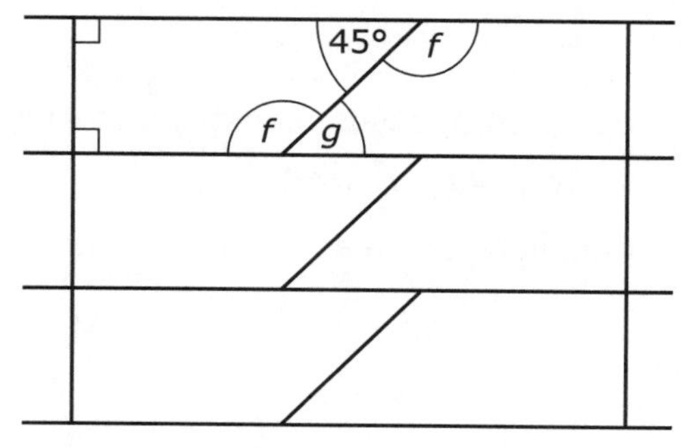

S1.6HW Constructing perpendiculars

1 You want to swim across the pool from A to the side BC.

Copy the diagram and draw your path from A so that you swim the shortest distance possible.

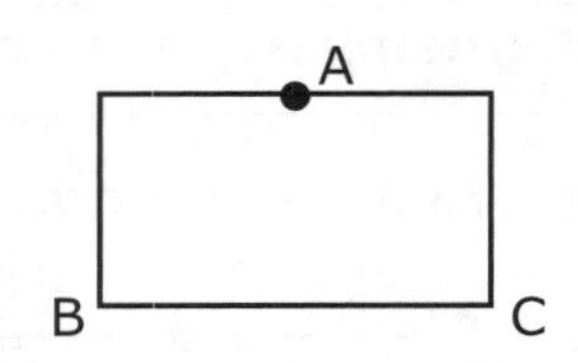

2 Draw a line BC so that BC = 10 cm.

Mark a point A about 5 cm from the line BC.

Construct the shortest line from A to the line BC.

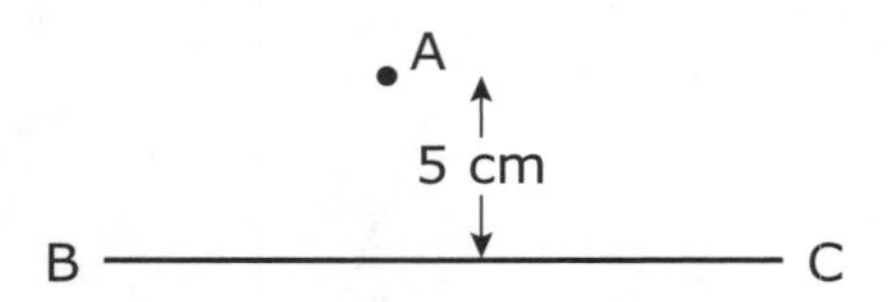

3 Draw a line AB so that AB = 10 cm.

Mark a point P so that AP = 4 cm.

Construct the perpendicular from the point P.

4 **a** Construct this kite using compasses.

b Calculate the area of the kite.

Hint: Draw DB first and then construct AC.

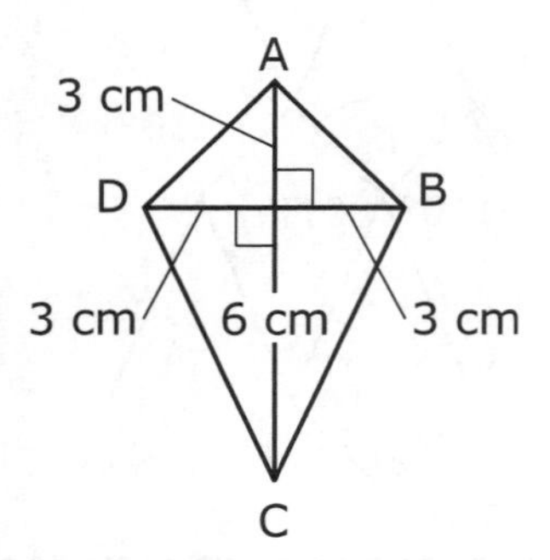

Hint:

Area of a triangle = $\frac{1}{2}hb$

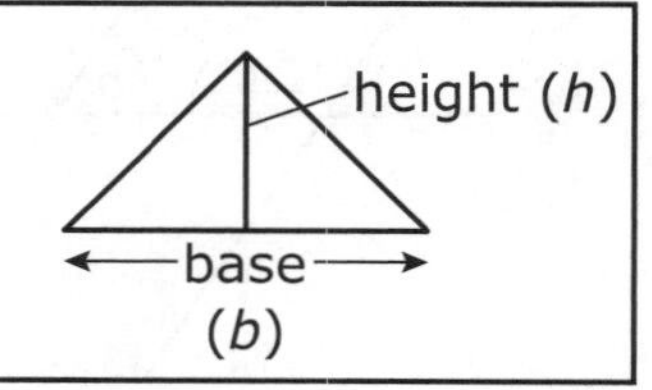

5 **a** Using compasses, construct the triangle ABC.

b Write the type of triangle.

c Calculate or measure the angles at A and C.

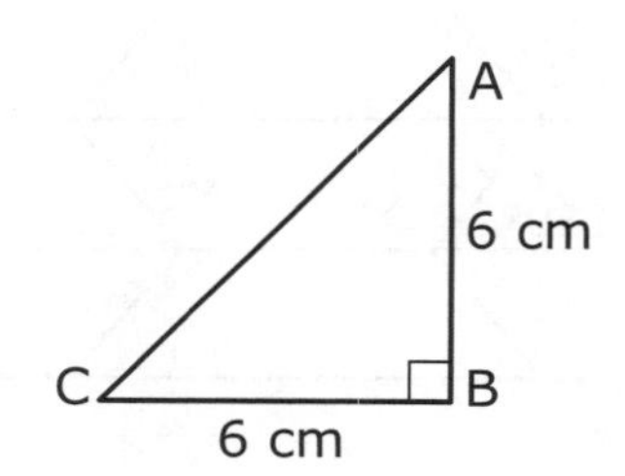

S1.7HW Constructing triangles

1 Accurately construct these triangles. Give the type of triangle for each one.

a ASA

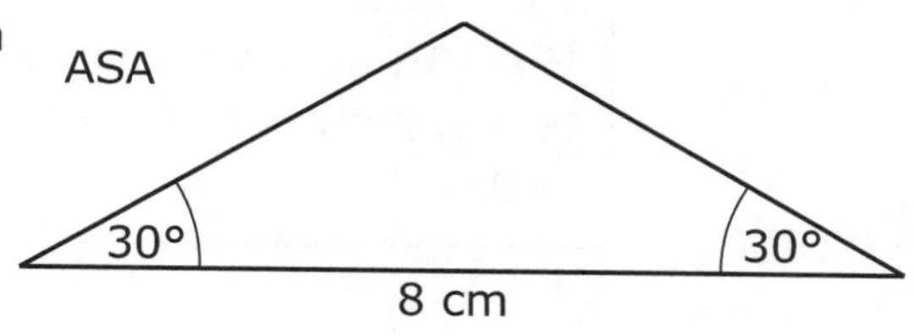

b ASA

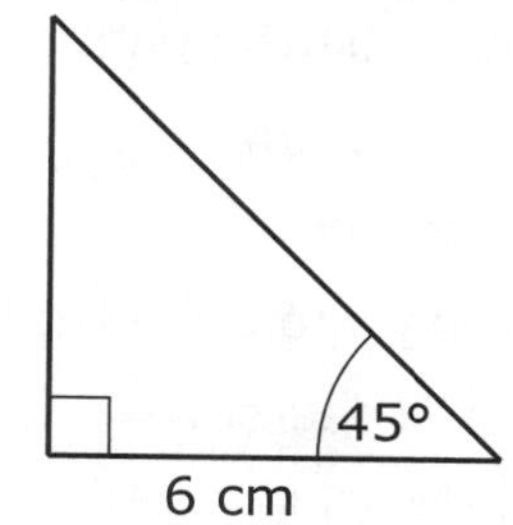

c SAS

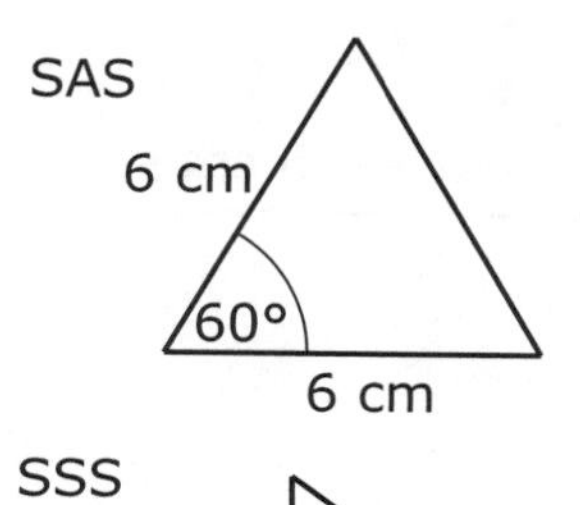

d SAS

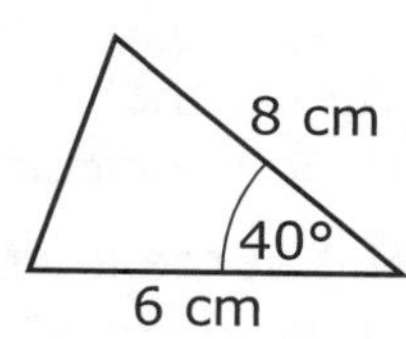

e SSS

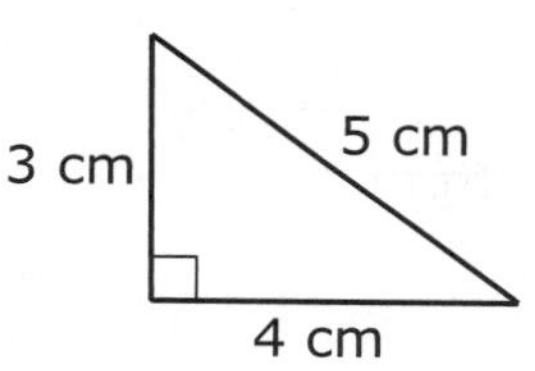

f SSS

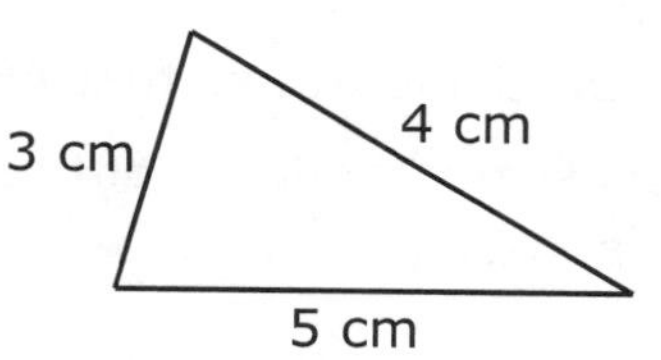

2 Accurately construct this net of a squared-based pyramid.

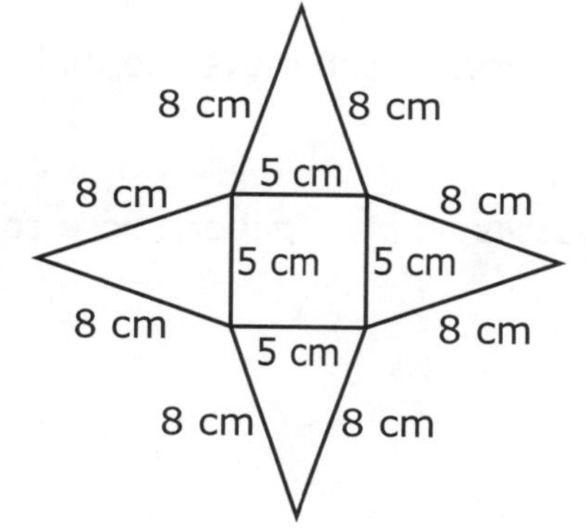

3 Accurately construct these quadrilaterals.

a

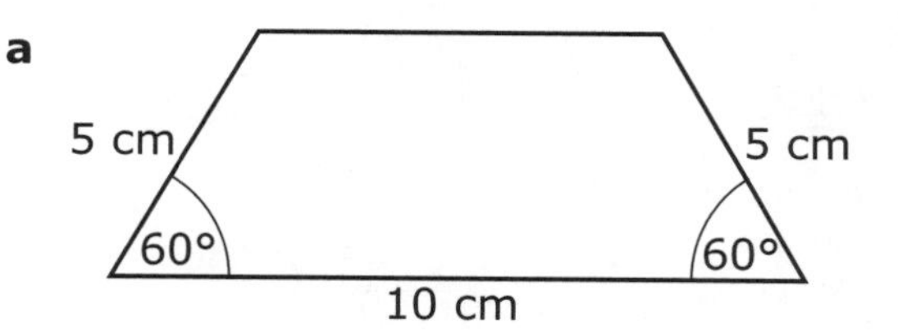

b

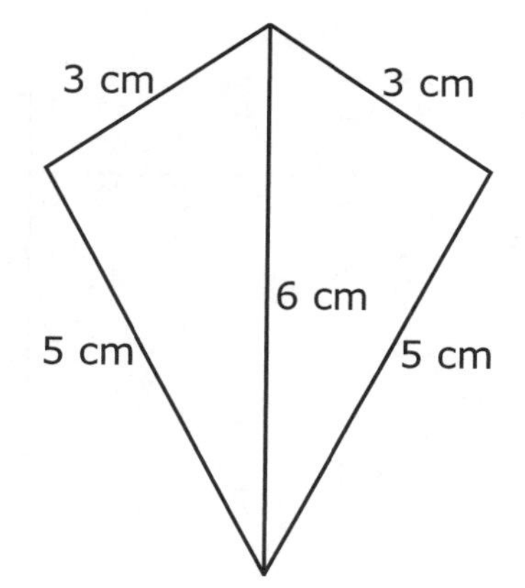

1 Draw the locus of each situation.

If possible, describe the path using mathematical terms.

a a clock pendulum

b a roller coaster

c a ball going into a basket in basketball

d a child on a see-saw

e a child on a slide

> **Hint:** Using mathematical terms, path **a** is an arc of a circle.

2 **a** Draw a line AB, so that AB = 5 cm.
Construct the perpendicular bisector of AB.

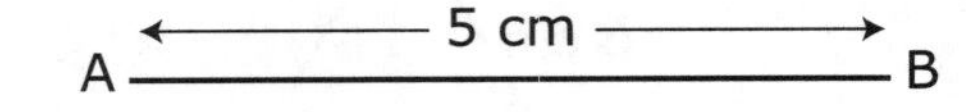

b Label the point at the intersection of AB and the bisector as M. Measure AM.

3 **a** Draw a line AB, so that AB = 7 cm.

b Construct the locus of the points that are equidistant from A and B.

4 Draw a point O.

Draw the locus of the point X, which moves so that the distance OX is always 4 cm.

5 Construct, using compasses on plain paper, these coordinate axes:

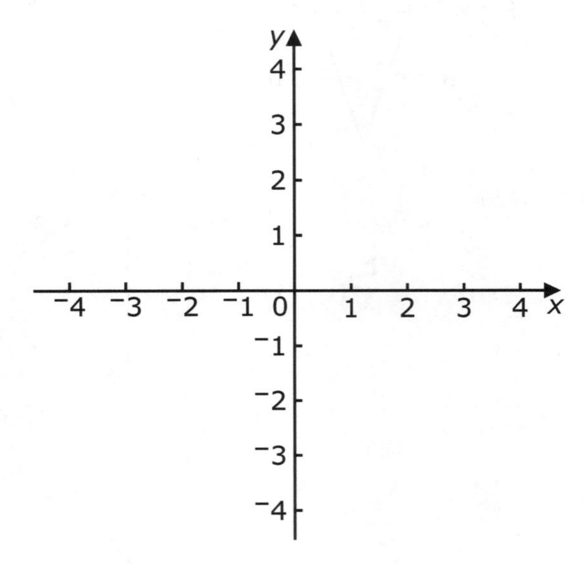

S1.9HW More loci

1 **a** Use a protractor to draw an angle of 70°.

b Construct the angle bisector.
Measure and label the two new angles.

2 **a** Draw vertical and horizontal lines that intersect at right angles.

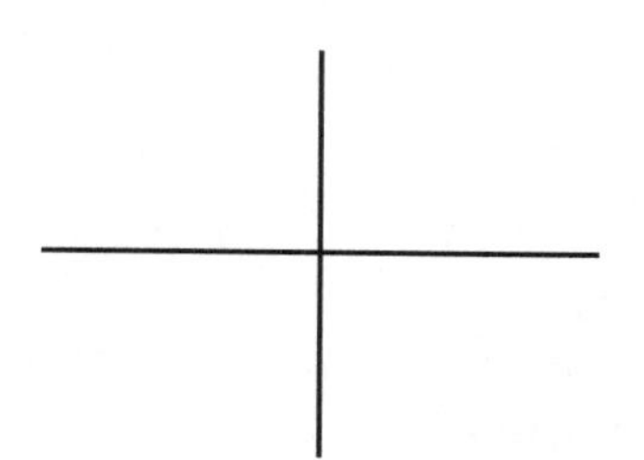

b Construct the angle bisector for each angle.

c Label your diagram N, NE, E, SE, S, SW, W, NW in a clockwise direction.

3 **a** Use a protractor to draw an angle of 120°.

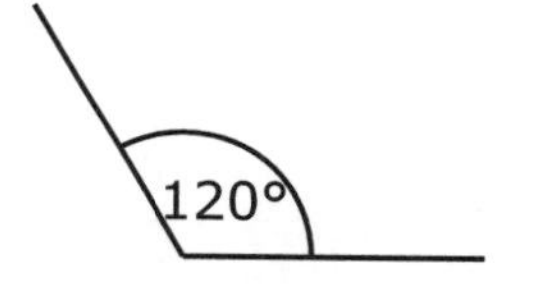

b Construct the angle bisector. Measure and label the two new angles.

Remember:

- In LOGO you specify distance in mm and angles in degrees.
- FORWARD 60 means forward 60 mm or 6 cm.
- LEFT 90 means turn left through 90°.

4 Write LOGO commands to draw these shapes.

a

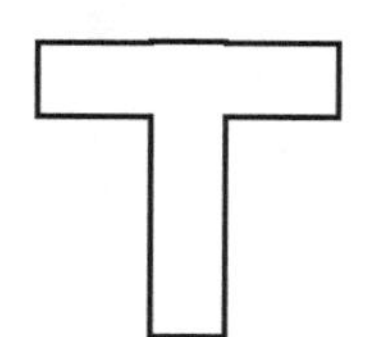

b

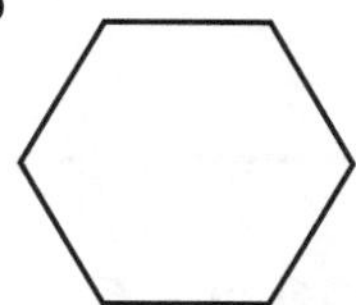

c

5 Draw a line AB, with AB = 8 cm.

Draw the locus of the points that are always 2 cm away from the line.

Level 4

Copy this rectangle twice.

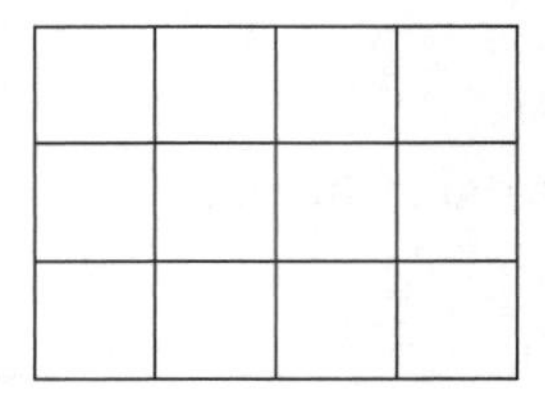

You can draw one straight line on the rectangle to make **two triangles.**

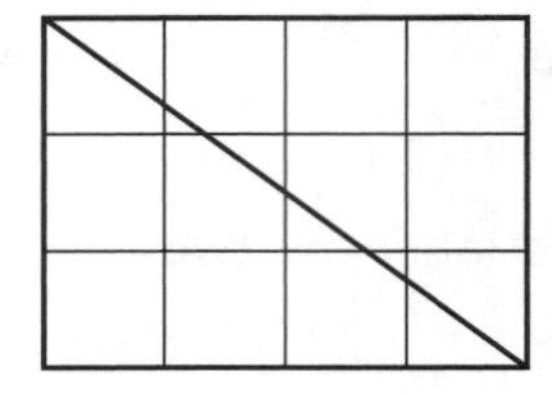

a Draw one straight line on one of your rectangles to make **one square** and **one rectangle** that is not a square. *1 mark*

b Draw one straight line on your other rectangle to make **one triangle** and **one quadrilateral**. *1 mark*

Level 5

Use compasses to construct a triangle that has sides **8 cm**, **6 cm** and **7 cm**.

Leave in your construction lines. *2 marks*

D1.1HW The handling data cycle

1 What is primary data?

Give an example of primary data.

2 What is secondary data?

Give an example of secondary data.

3 Here are some ideas for handling data projects.

Suggest a set of data that could be included in each project, and state whether it is primary or secondary data.

Explain how many data values you would expect there to be in each set.

a Jan wanted to investigate what temperature people let their tea cool down to before they started drinking it.

b Solomon wanted to know whether the race times in his school's sports days had improved over the years.

c Funda decided to see whether people could predict whether a coin would land on heads or tails.

d Siân wanted to know what were the most popular holiday destinations for people in the UK, and how many people in her school had visited them.

e Peter decided to find out what percentage of Year 9 students had a television in their bedroom.

D1.2HW Calculating statistics

1 Carl looked at this set of data.

Computer network performance – number of errors in a 30-day period

Results for 15 different computer networks

0	38	42	0	104
92	61	57	45	73
88	63	55	51	62

Carl said: 'The easiest average to work out is the mode, which is 0 errors.'

Explain why the mode would **not** be a good average to use in this case.

2 Zoë compared the prices of five second-hand cars advertised in her local newspaper.

	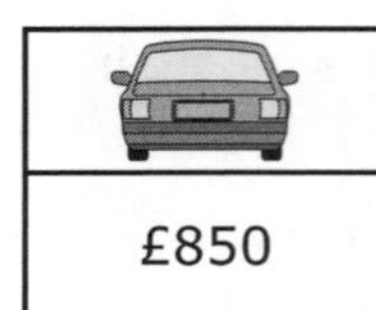	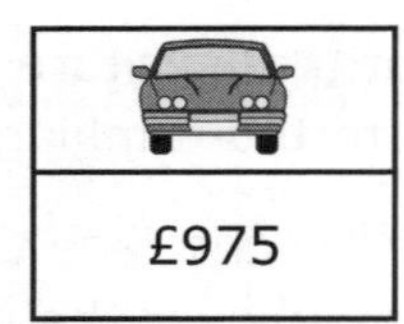	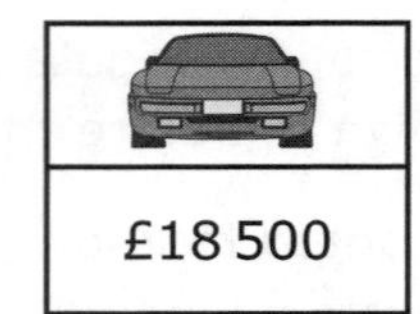	
£1200	£850	£975	£18 500	£695

a Explain why you could not find the mode of these prices.

b Explain why the mean would not give a very representative impression of the prices of the cars.

c Work out the median and the range of the prices.

3 Ronnie measured the heights of 20 people in her class.

Heights of 20 people in centimetres

148	162	157	172	168	176	166	170	180	179
160	173	175	178	153	182	176	185	165	169

Copy and complete the frequency table for this set of heights.

Use your completed table to work out the modal class for the heights.

Height, h cm	**Frequency**
$140 \le h < 150$	
$150 \le h < 160$	
$160 \le h < 170$	
$170 \le h < 180$	
$180 \le h < 190$	

Constructing pie charts

Remember:

- There are 360° in a circle.
- To find the angles in a pie chart:
  - Work out the total number of items.
  - Work out the angle for one item.
  - Multiply the angle for one item by the number of items in each category in turn.

1 The table shows the numbers of drinks dispensed by a vending machine one lunchtime.

Drink	Number
Tea	12
Coffee	9
Hot chocolate	10
Soup	5

Draw a pie chart to show this data.

2 A school secretary checked the number of packs of paper in the office stock cupboard. The table shows the number of packs of each colour.

Colour	Number
White	14
Pale blue	3
Dark blue	5
Yellow	7

Draw a pie chart for this data.

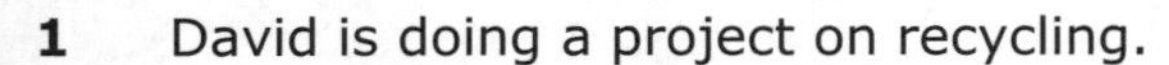

D1.4HW Selecting and drawing charts

1 David is doing a project on recycling.

He records the number of bottles put into a bottle bank in a half-hour period.

Bottle colour	Green	Brown	Clear	Blue
Number	15	12	9	2

Show this information as a bar chart.

2 This table shows the heights of trees in a small wood.

Height (m)	**Frequency**
$0 \le h < 5$	25
$5 \le h < 10$	30
$10 \le h < 15$	42
$15 \le h < 20$	38
$20 \le h < 25$	17

Draw a frequency diagram to represent this data.

Remember: A frequency diagram is like a bar chart, but it has no gaps between the bars.

D1.5HW Interpreting graphs and statistics

This population pyramid shows the age distribution of the population of Greathampton.

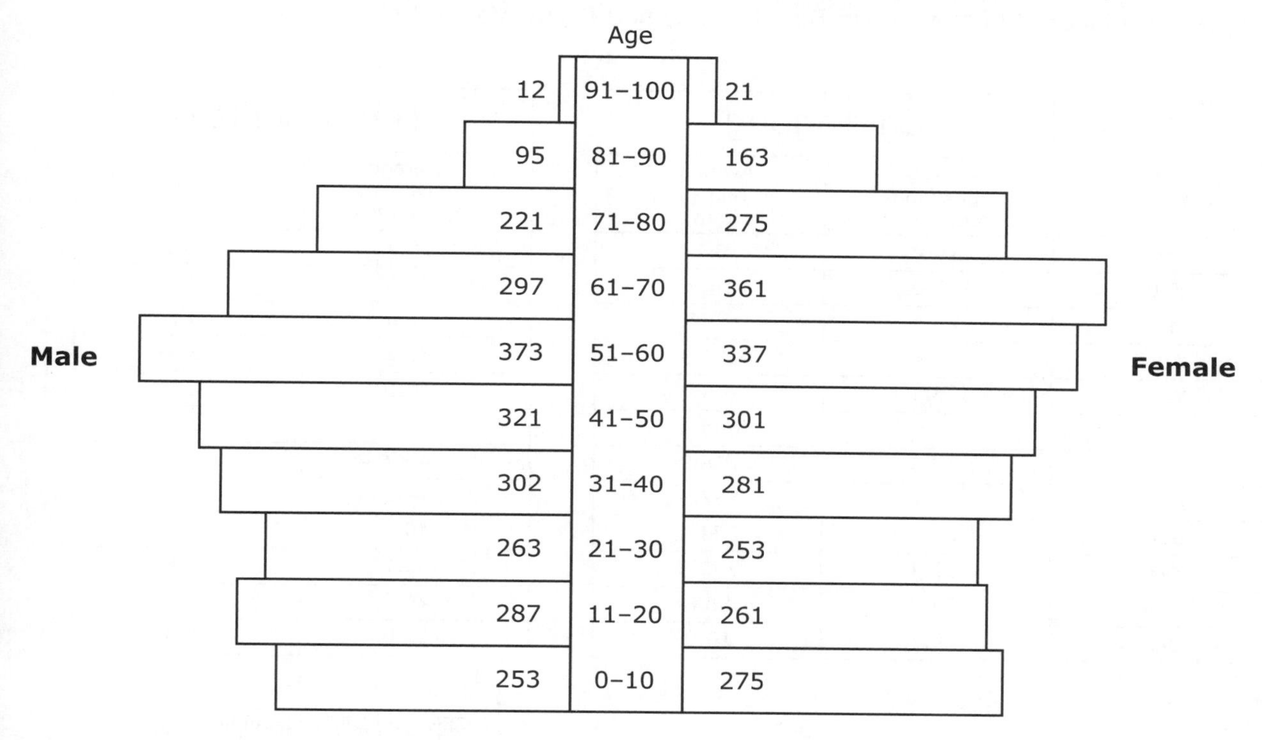

1 How many males are

a aged 40 or less

b aged between 21 and 60?

2 How many females are

a aged 30 or less

b aged between 51 and 80?

3 In Greathampton everyone over 60 is retired.

a How many people in Greathampton are retired?

b How many more females than males are retired?

4 What is the maximum possible number of teenagers in Greathampton?

5 Everyone aged 21–60 has a job. How many people have a job?

6 How many more people have a job than are retired?

D1.6HW Comparing distributions

The tables show the average monthly rainfall, and the average maximum temperatures, for two different places: Chittagong (in Bangladesh) and Death Valley (in the USA).

	Chittagong		Death Valley	
	Average maximum temp. (°C)	Average rainfall (mm)	Average maximum temp. (°C)	Average rainfall (mm)
Jan	19.9	6.7	18.2	8.2
Feb	22.0	14.8	22.6	12.3
Mar	25.6	53.6	26.4	8.7
Apr	27.6	116.3	31.1	3.7
May	28.4	246.7	36.6	2.1
Jun	27.9	603.7	42.2	1.0
Jul	27.6	718.9	45.7	3.2
Aug	27.7	552.9	44.5	2.8
Sep	27.9	284.4	40.2	4.2
Oct	27.4	242.5	33.1	2.8
Nov	24.2	58.8	24.2	5.6
Dec	20.7	10.0	17.6	4.3

1 Copy these axes on to graph paper.

Draw line graphs to show the average monthly temperatures in both places. You could draw both graphs on the same axes.

Temp °C
50
40
30
20
10
0
Jan Feb Mar Apr May Jun Jul Aug Sep Oct Nov Dec
Months

2 For each place work out:

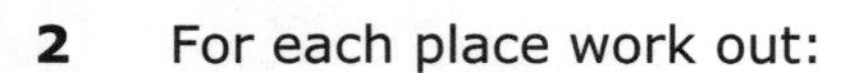

a the mean temperature

b the range of the temperature.

Write your answers in a table like this:

	Temperature	
	Mean	Range
Chittagong		
Death Valley		

3 For each place work out:

a the total rainfall **b** the mean rainfall **c** the range of the rainfall.

Write your answers in a table similar to the one in question **2**.

4 **a** Write a paragraph comparing the temperatures for the two places.

b Write a paragraph comparing the rainfall for the two places.

Level 4

There are 50 children altogether in a playgroup.

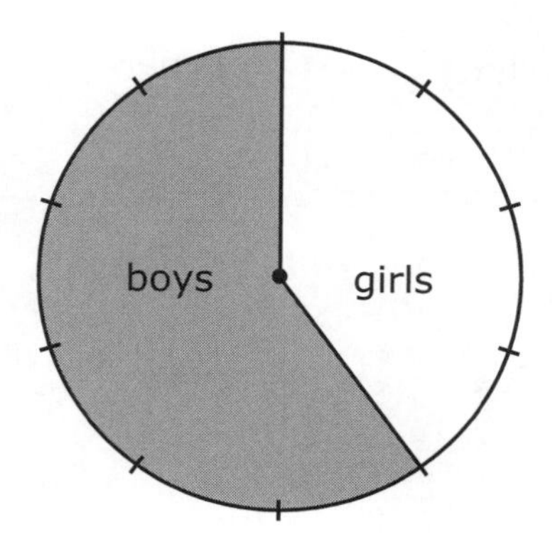

a i **How many** of the children are **girls**? *1 mark*

ii What **percentage** of the children are girls? *1 mark*

b **25** of the children are **4 years old**.

20 of the children are **3 years old**.

5 of the children are **2 years old**.

Show this information on a copy of the diagram below.
Label each part clearly.

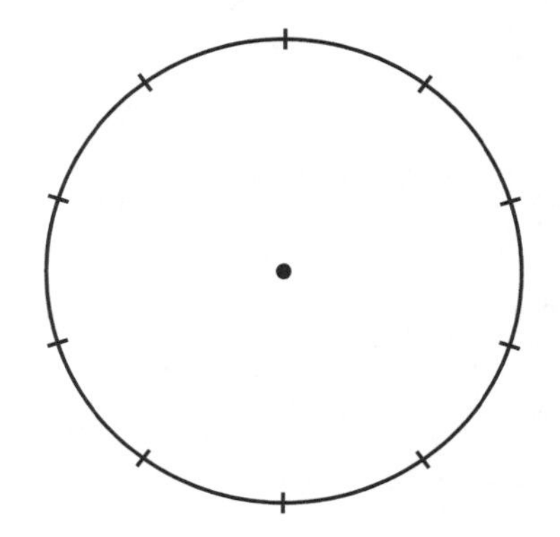

3 marks

Level 5

a There are four people in Sita's family.

Their shoe sizes are 4, 5, 7 and 10

What is the **median** shoe size in Sita's family? *1 mark*

b There are **three** people in John's family.

The **range** of their shoe sizes is **4**.

Two people in the family wear shoe size 6.
John's shoe size is **not 6** and it is **not 10**.

What is John's shoe size? *1 mark*

S2.1HW Coordinates

1 Copy these axes on to squared paper.
Plot and join up the points.
Write down the letter formed in each case.

a $(^{-}2, 2)\ (^{-}1, ^{-}2)\ (0, ^{-}2)$

b $(^{-}2, ^{-}2)\ (^{-}2, 2)\ (0, 0)\ (2, 2)\ (2, ^{-}2)$

c $(^{-}2, ^{-}2)\ (0, 2)\ (2, ^{-}2)\ (1, 0)\ (^{-}1, 0)$

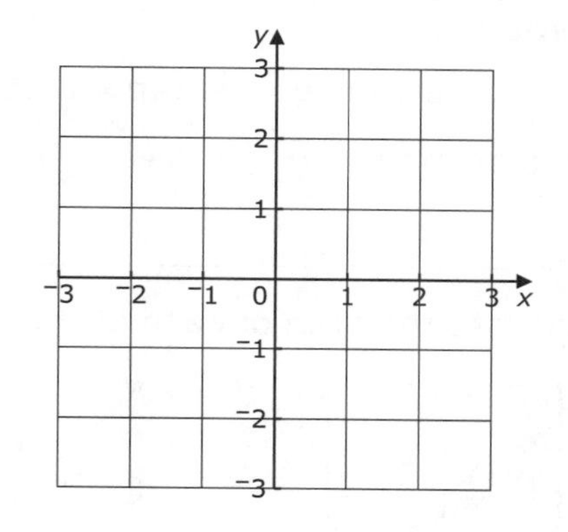

2 Copy these axes and plot the points shown.
Write down, if possible, the coordinates of the extra point that will make:

a a parallelogram

b an isosceles trapezium

c a kite

d a rhombus

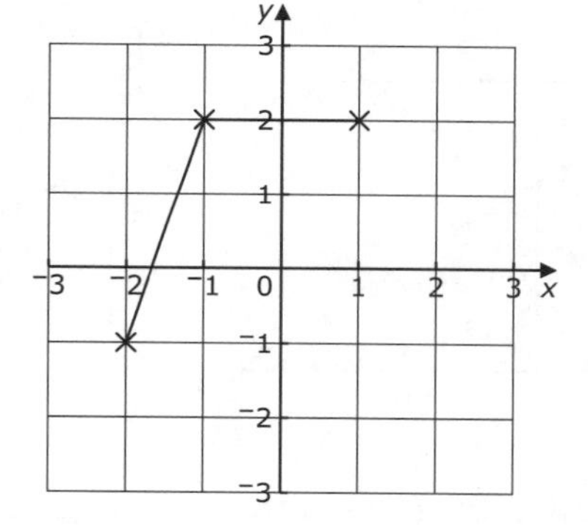

3 Copy these axes. Plot the pairs of points and find the midpoints of the lines joining them.

a (6, 0) and (6, 4)

b (2, 6) and (6, 6)

c (0, 3) and (4, 5)

d (3, 4) and (5, 0)

e $(^{-}1, 1)$ and (3, 3)

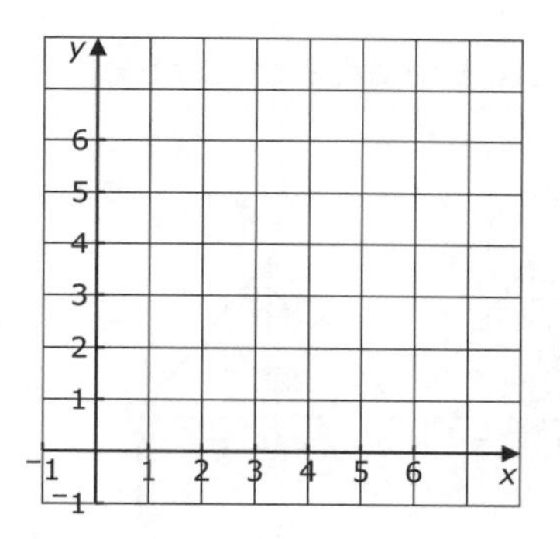

Remember:

- To find the midpoint of a line, you find the mean of the coordinates of the endpoints.
- The *x*-coordinate of the midpoint is the mean of the x-coordinates.
- The *y*-coordinate of the midpoint is the mean of the y-coordinates.

4 Check your answers to question **3** by finding the midpoints by calculation.

S2.2HW Area of a triangle

Remember:

- Area of rectangle = length × width
- Area of triangle = $\frac{1}{2}$ × base × height

The height must be perpendicular to the base.

1 Calculate the area of each of these triangles.

a 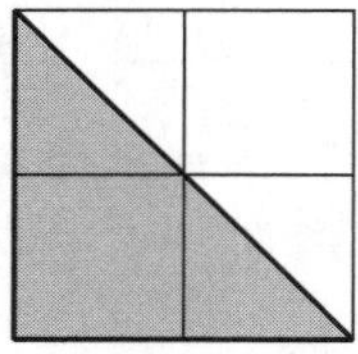

b 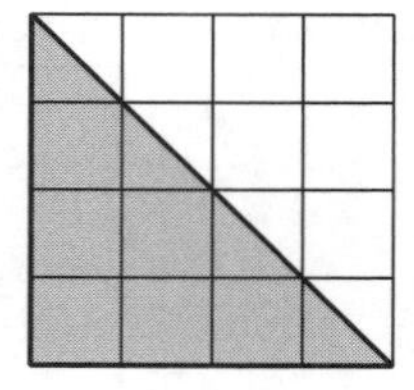

c 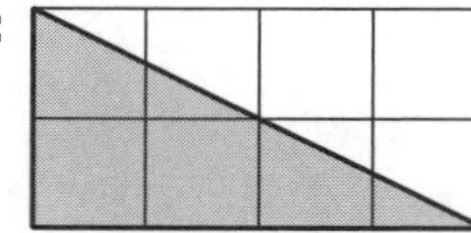

d 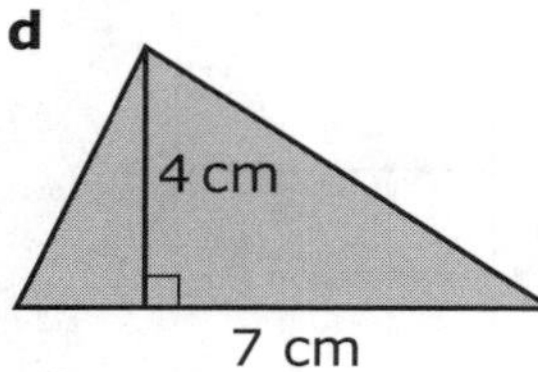

e 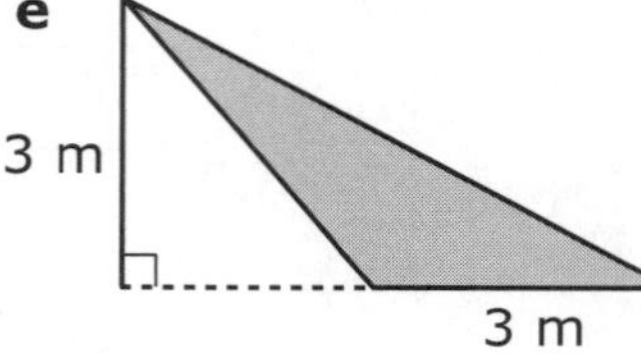

f 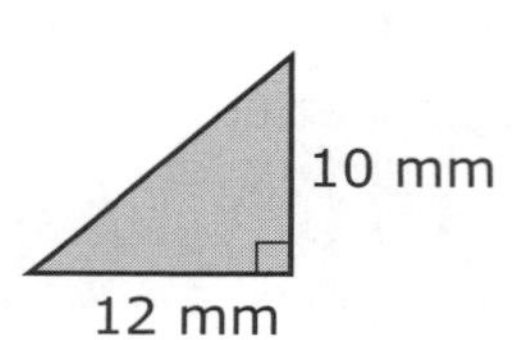

2 Draw a triangle with an area of 12 cm^2.

3 Calculate the area of each of these shapes.

a 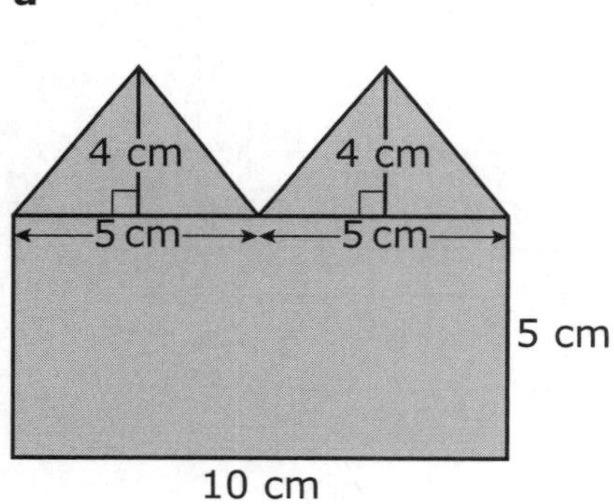

b 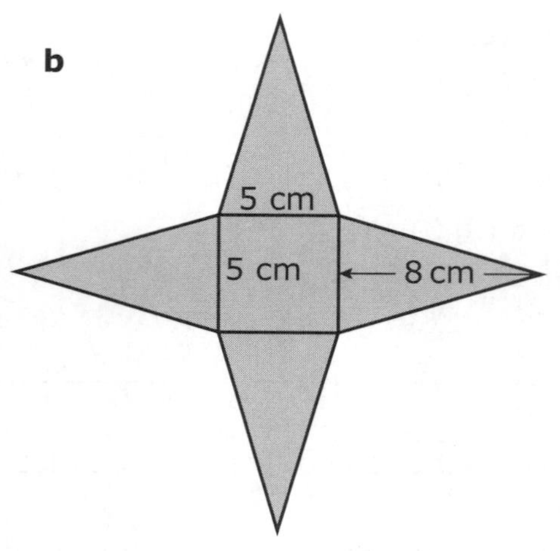

4 The area of this isosceles right-angled triangle is 8 cm^2.
Calculate the base and perpendicular height.

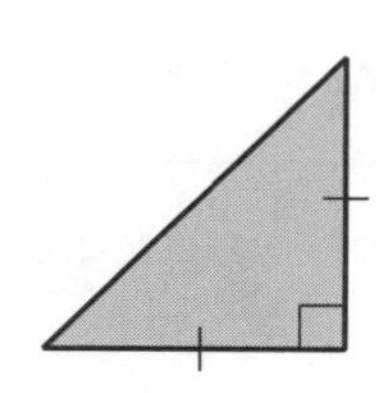

S2.3HW Area of a parallelogram and a trapezium

Remember:

- Area of triangle = $\frac{1}{2} \times$ base $\times$ height
- Area of parallelogram = base $\times$ height
- Area of trapezium = $\frac{1}{2} \times (a + b) + h$

The height must be perpendicular to the base.

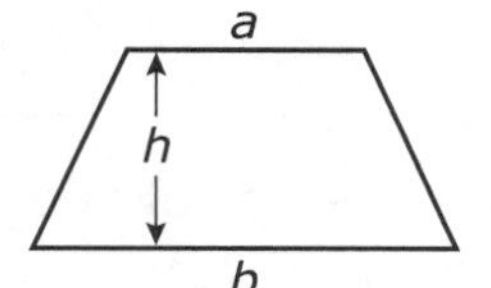

1 Calculate the area of each of these parallelograms.

a

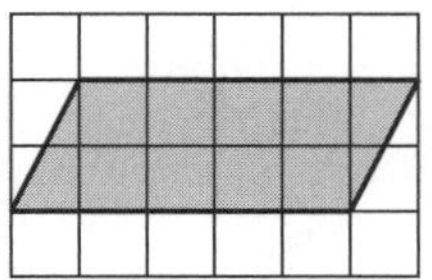

b

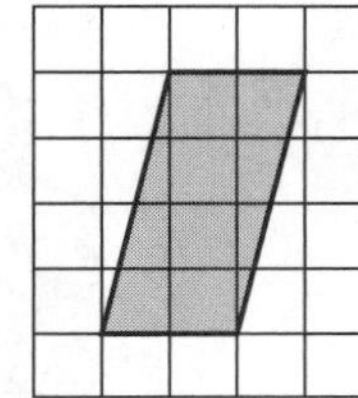

c

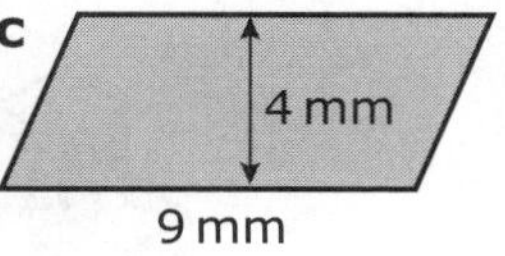

d

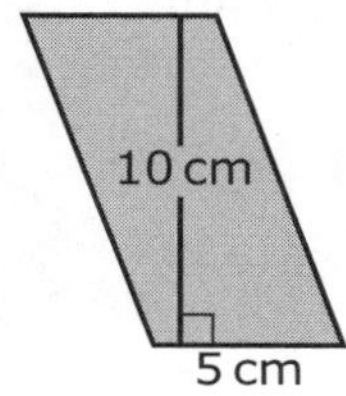

2 **a** Calculate the area of parallelogram A.

b Calculate the area of triangle B.

c Calculate the area of triangle C.

d Add your answers for the areas of A, B and C.

e Check your answer to **d** by calculating the area of the large triangle.

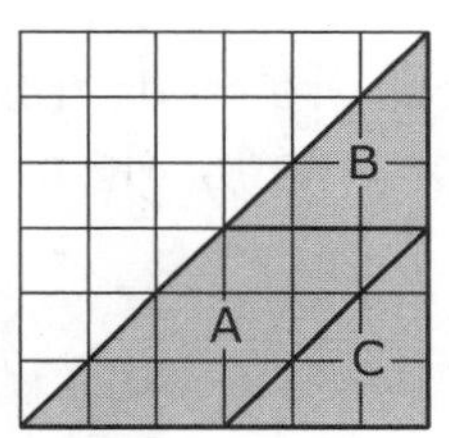

3 Calculate the area of each of these trapeziums.

a

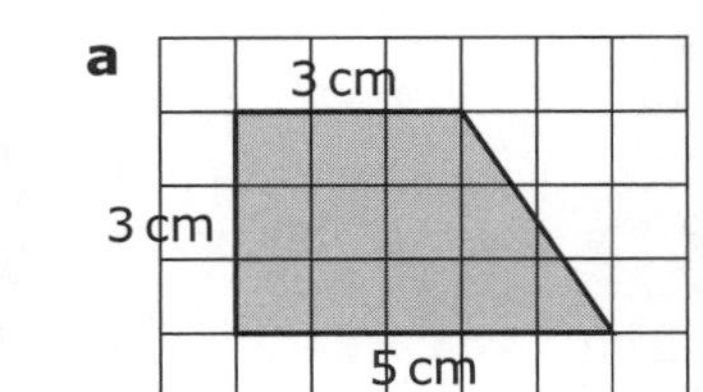

b

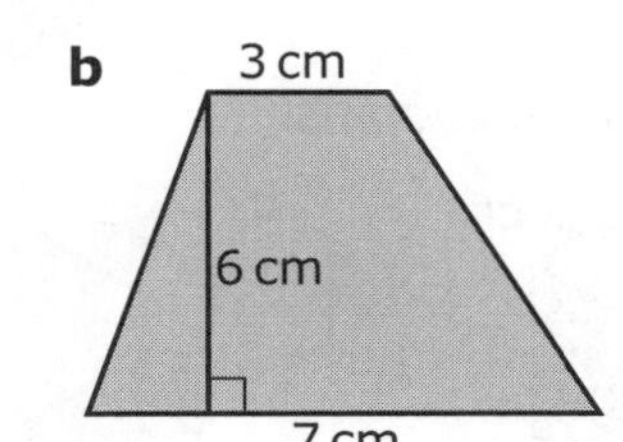

4 **a** Calculate the area of the trapezium using the trapezium formula.

b Calculate the area of triangle A.

c Calculate the area of parallelogram B.

d Add your answers for the areas of A and B. Check your answer – it should be the same as your answer to part **a**.

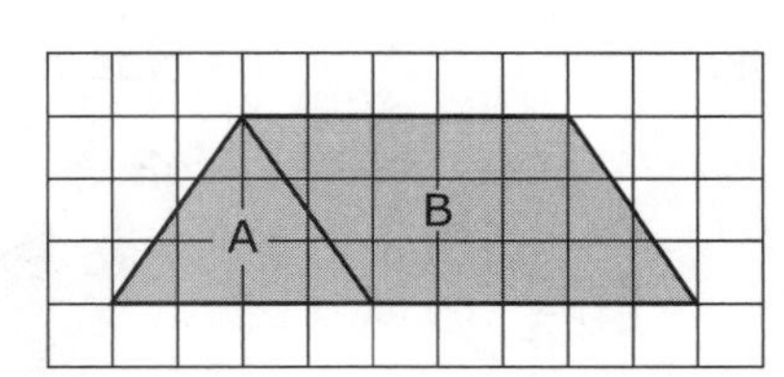

S2.4HW Measuring volume

Remember:

- Volume of cuboid = length × width × height

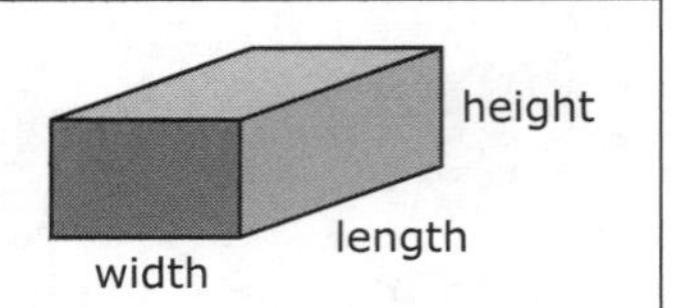

1 Give an object that would have its volume measured in:

a cm^3 **b** m^3

2 All these shapes are made from centimetre cubes.
Find the volume of each shape.

a 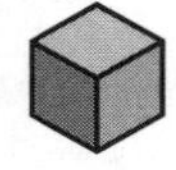

b 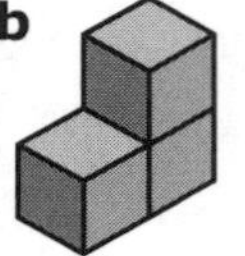

c 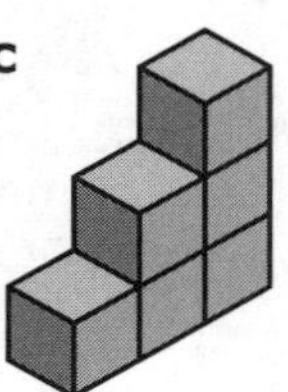

d 

e 

3 All these shapes are made from centimetre cubes.
Write down the length, width, height and volume for each cuboid.

a 

b 

c 

4 Calculate the volume of each of these cuboids.

a 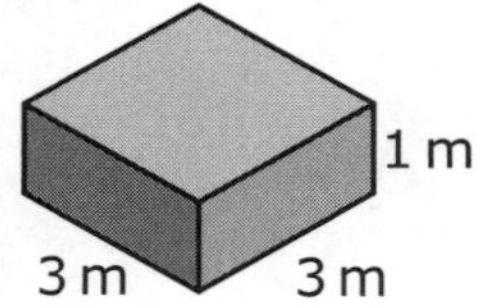

b 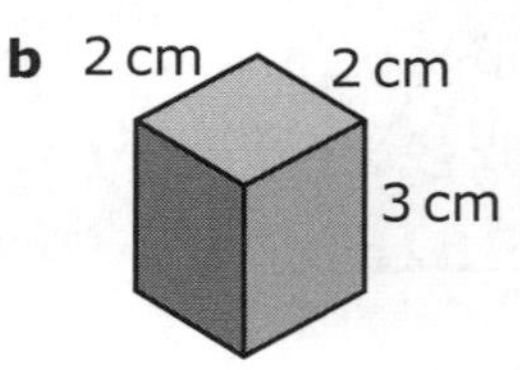

c 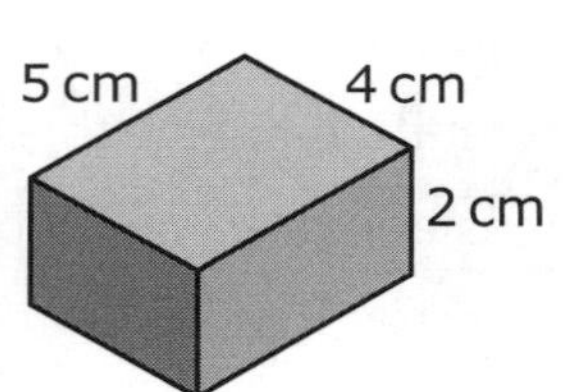

5 Find the length of the side of each of these cubes.

a 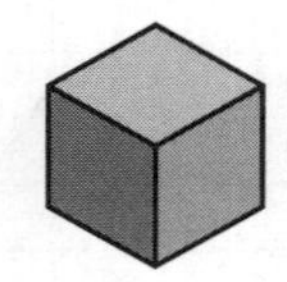

Volume = $27\ cm^3$

b 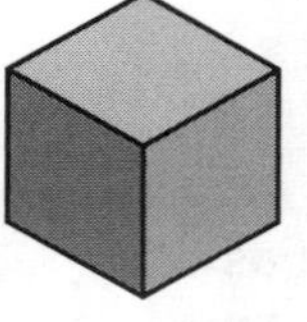

Volume = $1000\ cm^3$

c 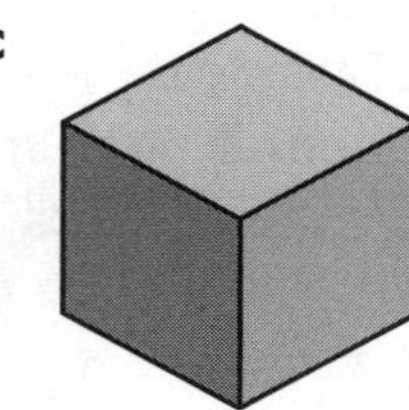

Volume = $3375\ cm^3$

S2.5HW Surface area

Remember:

- Volume of cuboid = length × width × height
- The surface area of a cuboid is the area of its net.

1 Copy these nets on to squared paper and cut them out.

a

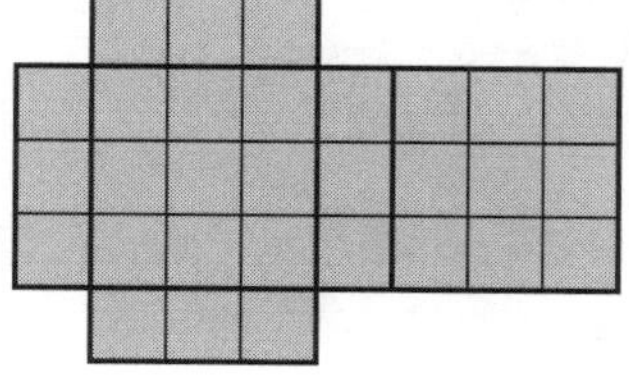

b

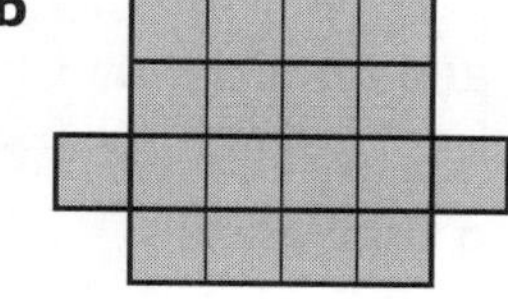

Fold each net into a 3-D shape.

For each shape find:

i the total surface area

ii the dimensions

iii the volume.

2 Calculate the total surface area and volume of each cuboid.

a

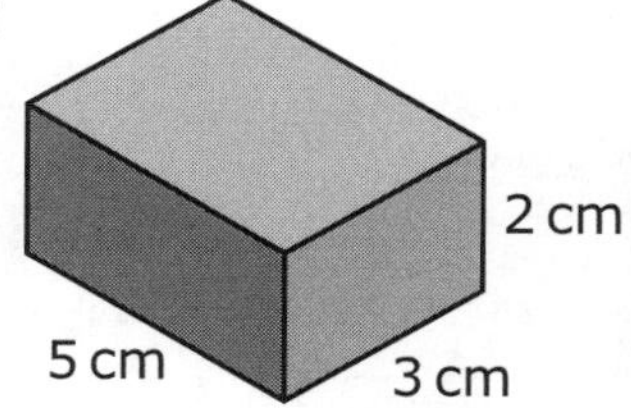

b

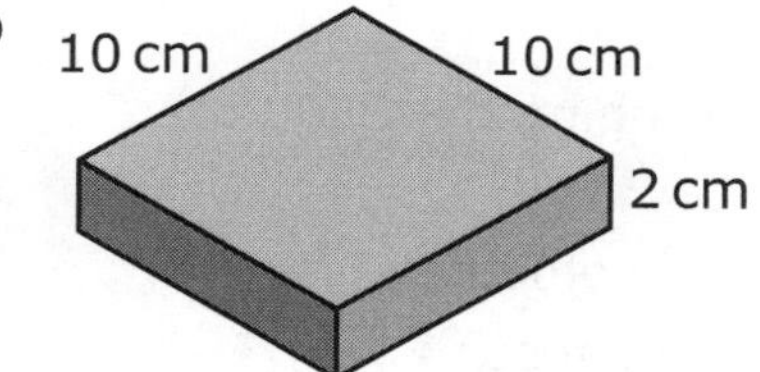

3 There are two different cuboids with a volume of 10 cm^3.

a Find the dimensions of each cuboid.

b Calculate the total surface area of each cuboid.

4 **a** Calculate the total surface area of this cube.

b Find the dimensions of the cube.

c Calculate the volume of the cube.

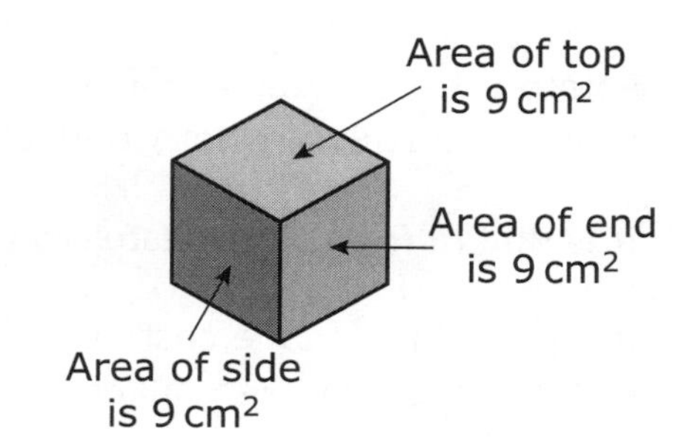

S2.6HW Units of measurement

Remember:

- 1 pint ≈ 600 ml
- 1 gallon ≈ $4\frac{1}{2}$ litres
- 1 yard ≈ 1 metre
- 5 miles ≈ 8 kilometres
- 1 foot ≈ 30 cm
- 1 kg ≈ 2.2 lb
- 1 yard = 3 feet

1 Suggest an appropriate metric and imperial unit to measure:

- **a** the height of a house
- **b** the capacity of a swimming pool
- **c** the mass of a car
- **d** the distance run in a marathon
- **e** the length of a swimming pool
- **f** your own mass.

2 Put each pair of quantities in order of size, smallest first.

- **a** 1 hour, 1 minute
- **b** 1 gram, 1 lb
- **c** 1 metre, 1 yard
- **d** 1 litre, 1 gallon
- **e** 1 metre, 1 foot

3 In 2003 the men's record for running the 1500 m was 3 minutes 26 seconds.

- **a** Change 1500 metres into kilometres.
- **b** Change the record time into seconds.

4 A box contains 750 g of cereal. Change this mass into kg.

5 A mug holds 250 ml of water.
How many mugs are needed to make one litre?

6 Potatoes used to be sold in lbs.
If one lb cost 60p, what would be the approximate cost of 1 kg?

7 It is 40 km from Manchester to Liverpool. What is this distance in miles?

8 If a pint of lemonade costs £3, what would you expect to pay for a litre of the same lemonade?

Level 4

a These cuboids are made from small cubes.

Write **how many small cubes** there are in each cuboid.

The first is done for you.

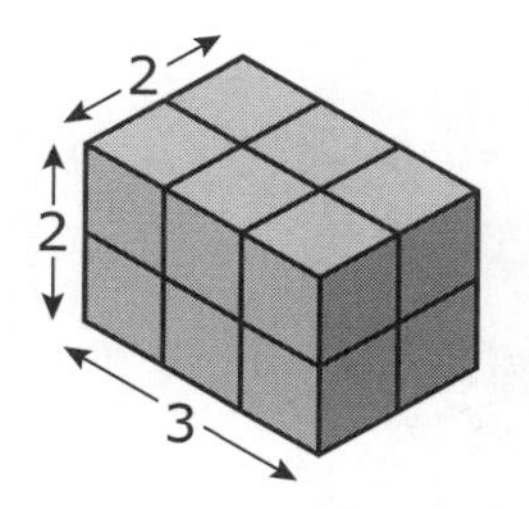

Number of cubes = 12

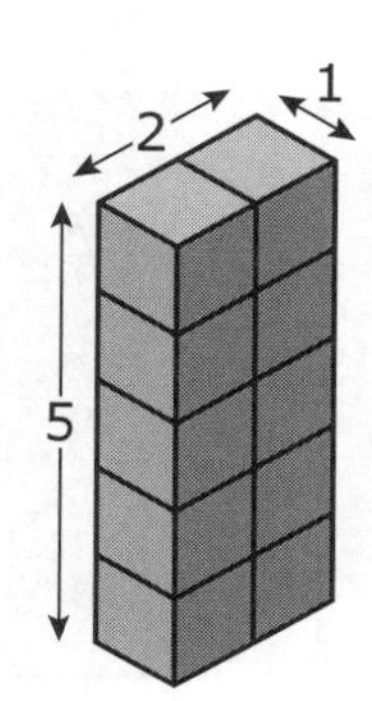

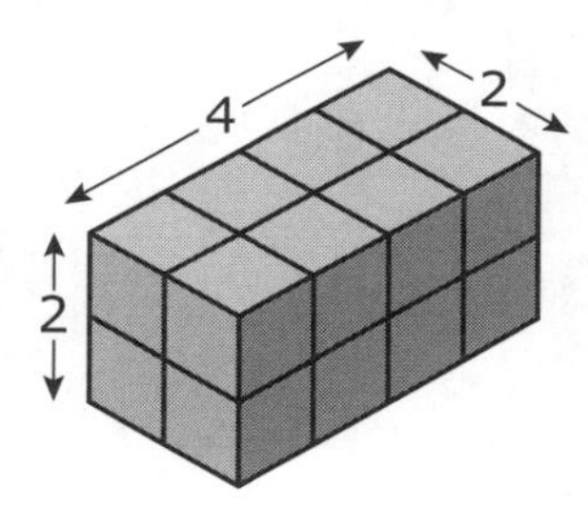

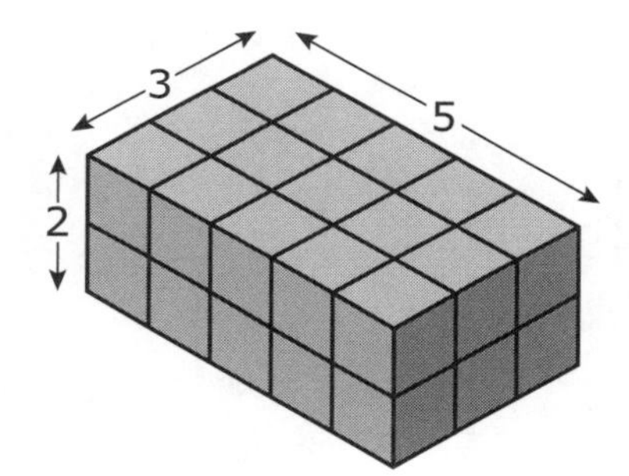

3 marks

b This shape is made with two cuboids.

Write **how many small cubes** there are in this shape.

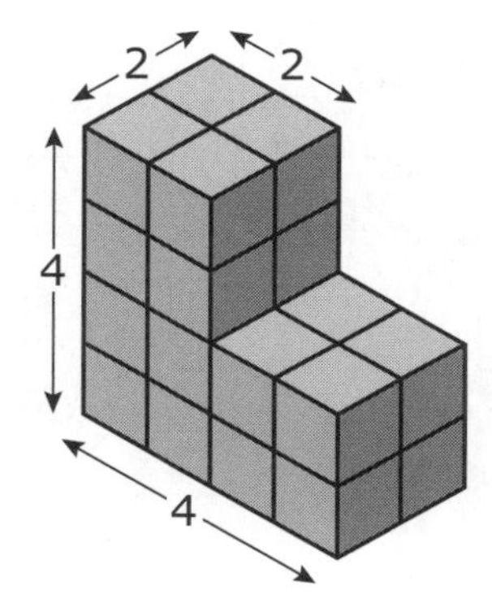

1 mark

Level 5

a List the rectangles that have an area of 12 cm^2.

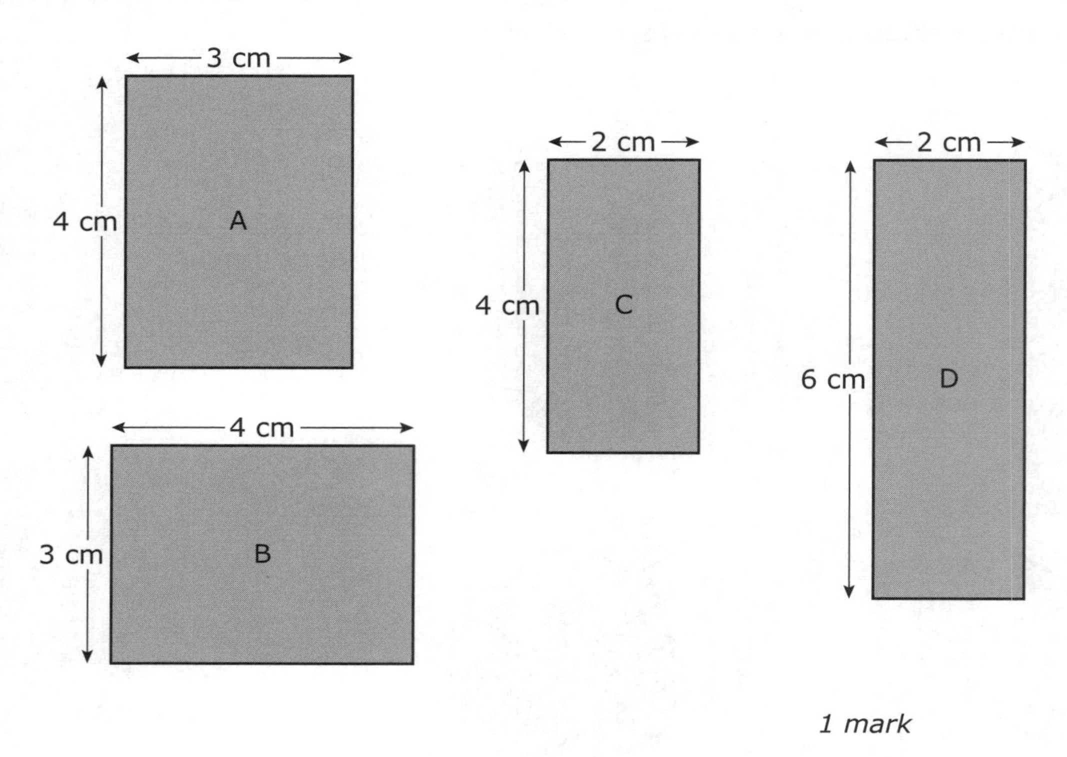

1 mark

b A **square** has an area of **100** cm^2.

What is its **perimeter**?

Show your working.

2 marks

N2.1HW Powers and roots

1 Work out the answers to:

a 5^2 **b** $\sqrt{36}$ **c** 1^2

d 3^3 **e** $\sqrt{16}$ **f** 7^2

g $\sqrt{4}$ **h** $\sqrt[3]{125}$ **i** $\sqrt{64}$

j $\sqrt[3]{64}$ **k** $\sqrt[3]{1000}$ **l** 9^2

Hints:
$5^2 = 5 \times 5$
$\square \times \square = 36$
$\sqrt{36} = \square$
$3^3 = 3 \times 3 \times 3$
$\square \times \square \times \square = 125$
$\sqrt[3]{125} = \square$

2 Use your answers to question **1** to help you work out:

a 50^2 **b** $\sqrt{3600}$ **c** 100^2

d 30^3 **e** $\sqrt{1600}$ **f** 70^2

g $\sqrt{400}$ **h** $\sqrt[3]{125\,000}$ **i** $\sqrt{6400}$

j $\sqrt[3]{64\,000}$ **k** $\sqrt[3]{1\,000\,000\,000}$ **l** 900^2

3 A well-known square number fact is that $3^2 + 4^2 = 5^2$.

Copy and complete these square number facts.

a $6^2 + \square^2 = 10^2$

b $\square^2 + 12^2 = 13^2$

4 In question **3** the number fact was $a^2 + b^2 = c^2$.

However, there are many square numbers, that add up to a cube number, for example $a^2 + b^2 = c^3$.

Copy and complete these.

a $5^2 + 10^2 = \square^3$

b $2^2 + \square^2 = 2^3$

c $\square^2 + 30^2 = 10^3$

d $2^2 + \square^2 = 5^3$

N2.2HW Powers of 10

1 Copy this number ladder with 15 spaces.

Now work out the answers to each of these questions and write the answers in the correct order on a copy of the number ladder.

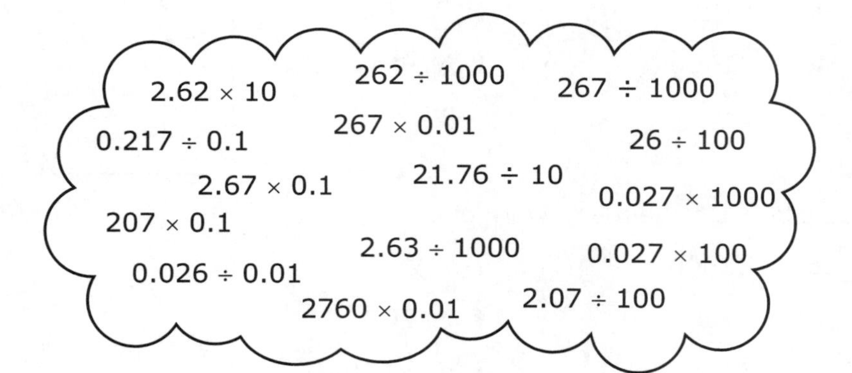

largest
smallest

2 Now make up your own ten questions to fill a 10-space number ladder.

Write your questions and their answers in a copy of the number ladder in the correct order.

largest
smallest

N2.3HW Rounding

A recurring decimal goes on forever. You put a dot over the start of the numbers that repeat.

Using a calculator, $\frac{1}{11} = 1 \div 11 = 0.090909090...$

This can be written as $0.\dot{0}\dot{9}$ as 09 recurs.

1 Copy and complete this table.

Fraction	Calculator display	Recurring notation	Rounded to	
			1 dp	2 dp
$\frac{1}{11}$	0.090909090	$0.\dot{0}\dot{9}$	0.1	0.10
$\frac{2}{11}$	0.181818181	$0.\dot{1}\dot{8}$		
$\frac{3}{11}$				
$\frac{7}{11}$				
$\frac{1}{6}$	0.166666666	$0.1\dot{6}$	0.2	0.17
$\frac{2}{6}$				
$\frac{4}{6}$				
$\frac{5}{6}$				
$\frac{2}{7}$	0.285714285	$0.\dot{2}8571\dot{4}$		
$\frac{3}{7}$				
$\frac{5}{7}$				

2 Copy this table.

Use a calculator to find five recurring decimals of your own.

Record the information in the table.

Fraction	Calculator display	Recurring notation	Rounded to	
			1 dp	2 dp

N2.4HW Addition and subtraction

Number of visitors to a stately home each month

Month	Adult visitors	Child visitors
Jan	46	7
Feb	102	25
Mar	98	14
Apr	149	97
May	176	84
Jun	213	124
Jul	241	145
Aug	276	152
Sep	193	62
Oct	139	74
Nov	59	21
Dec	37	9

1 How many people visited in April and October, in total?

Hint: 'People' includes adults and children.

2 Calculate the total number of adult visitors for the summer months: June, July and August.

3 Calculate the total number of child visitors for the winter months: December, January and February.

4 How many more adult visitors were there in September than in October?

5 How many more child visitors were there in July than in December?

6 Which month had most visitors? How many visitors was this?

Hint: Include adults and children

7 Which month had least visitors? How many visitors was this?

N2.5HW Mental multiplication and division

It is easy to double: $3 \times 2 = 6$

Multiplying by 4 is easy too: $4 = 2 \times 2$

$$6 \times 4 = 6 \times 2 \times 2$$
$$= 12 \times 2$$
$$= 24$$

1 Copy and complete these prime factor tree diagrams.

a

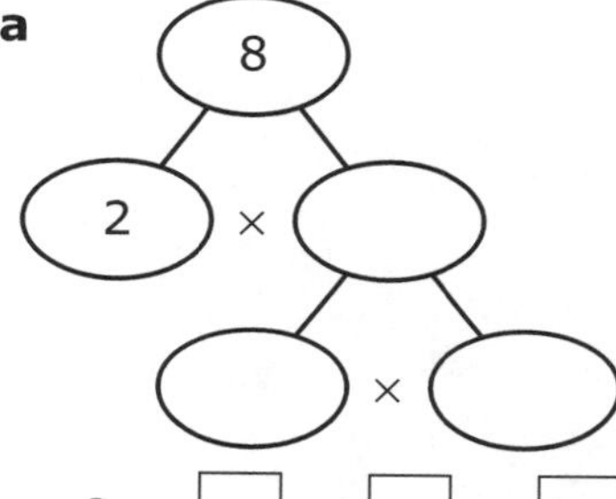

8 = □ × □ × □

b

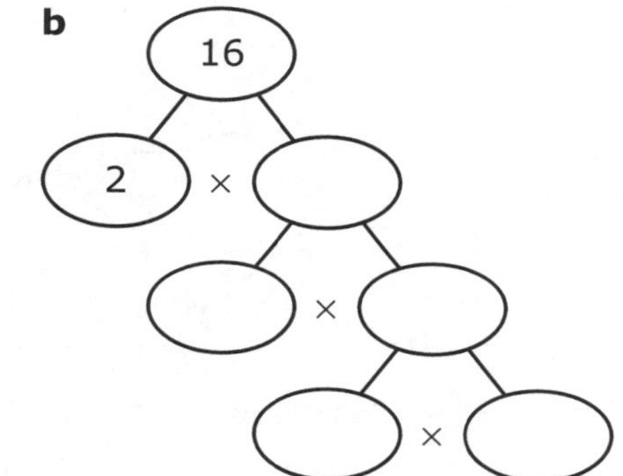

16 = □ × □ × □ × □

2 Use your diagrams from question **1** to help you work out:

a 12×8 **b** 16×12

c 15×8 **d** 0.15×8

e 22×0.8 **f** 32×16

g 1.6×24 **h** 0.16×2.4

3 Now adapt your method to help you work out:

a $280 \div 8$ **b** $208 \div 16$

c $7.2 \div 8$ **d** $41.6 \div 1.6$

N2.6HW Written multiplication

Remember:

- For the calculation 193 × 179

  Last digit check
  It ends in 3 × 9 so the answer will end in a 7.

  Estimation
  It is roughly 200 × 200 = 400

Hint:
3 × 9 = 27

1 Look at these numbers.

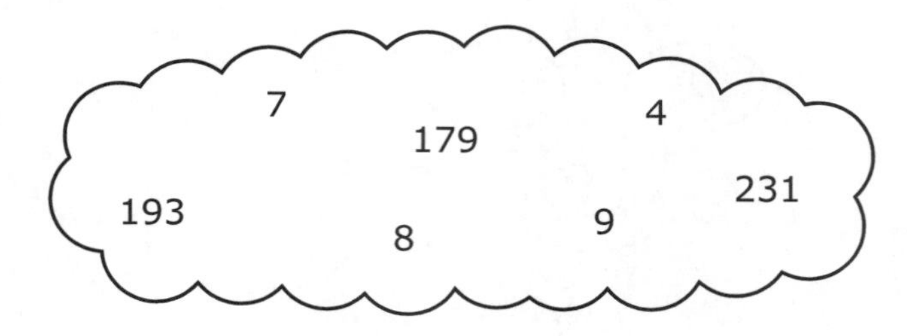

Three of them multiply together like this:

_______ × _______ × _______ = 14 553

Identify the three correct numbers using written calculations.
Use the methods of **last digit check** and **estimation** to help you.

2 In her homework Shani calculates:

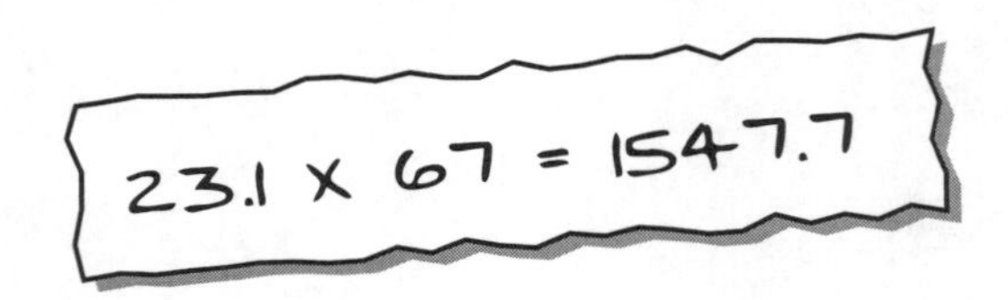

a Do a written calculation to check whether Shani's answer is correct.

b Write down four other calculations that equal 1547.7.
Here is one: 2.31 × 670.

Written division

Example

To work out $1204 \div 22$

Estimate: $1200 \div 20 = 60$

Work out roughly what the answer is:

$50 \times 22 = 1100$

$60 \times 22 = 1320$

1204

1100 1320

The answer is between 50 and 60.

$50 \times 22 = 1100$, so subtract this from 1204 to find what's left:

$1204 - 1100 = 104$

Write out the 22 times table:

$1 \times 22 = 22$ $2 \times 22 = 44$ $3 \times 22 = 66$ $4 \times 22 = 88$ $5 \times 22 = 110$

$1204 = 50 \times 22 + 4 \times 22 + 16$

$= 54 \times 22 + 16$

$1204 \div 22 = 54$ remainder 16

Here are 10 division questions.

i	$1034 \div 22$	**ii**	$842 \div 15$
iii	$646 \div 19$	**iv**	$861 \div 21$
v	$840 \div 24$	**vi**	$962 \div 13$
vii	$579 \div 34$	**viii**	$929 \div 32$
ix	$1189 \div 41$	**x**	$1089 \div 33$

1 Without doing the calculations, identify those that will leave a remainder.

Explain how you know.

Hint: Do not do the division.

2 Choose five of the questions and work them out using a written method.

Remember to do an estimate first.

3 Which of the 10 numbers being divided is a square number?

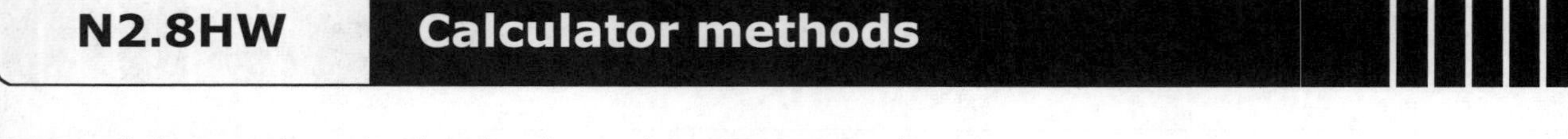

N2.8HW Calculator methods

Some of the numbers in the cloud are the answers to the questions below.

A number may be used more than once!

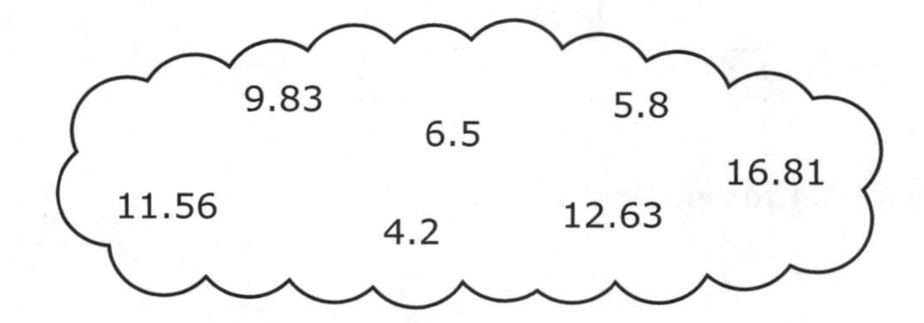

You can use a calculator to help you.

1 Which number added to 37.17 makes 49.8?

2 Which two numbers have a product of 82.095?

3 What is 100.34 divided by to make 17.3?

4 The difference between which two numbers is 4.03?

5 Which number is the square root of 96.6289?

6 Which number is missing from this calculation?

$38.6 - 6.4 \times \square = 11.72$

7 What are the two numbers missing in this calculation?

$\square^2 + \square^2 = 114.2689$

8 Which two numbers have square roots that sum to 7.5?

Level 4

a Claire puts a **2** digit whole number into her calculator.

She **multiplies** the number by **10**.

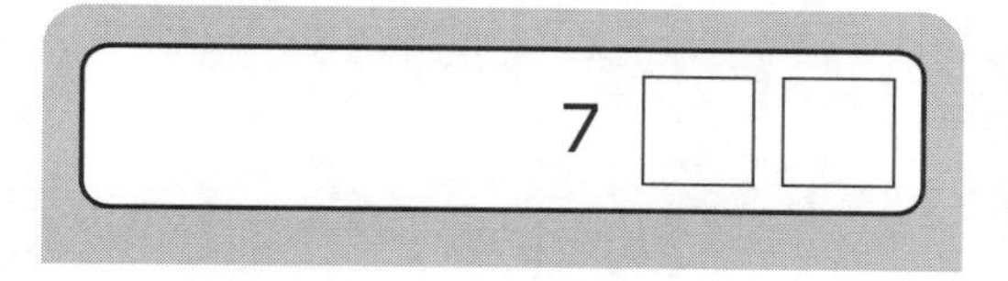

Write **one** other digit which you **know** must be on the calculator display.

1 mark

b Claire starts again with the **same** 2 digit whole number.

This time she **multiplies** it by **100**.

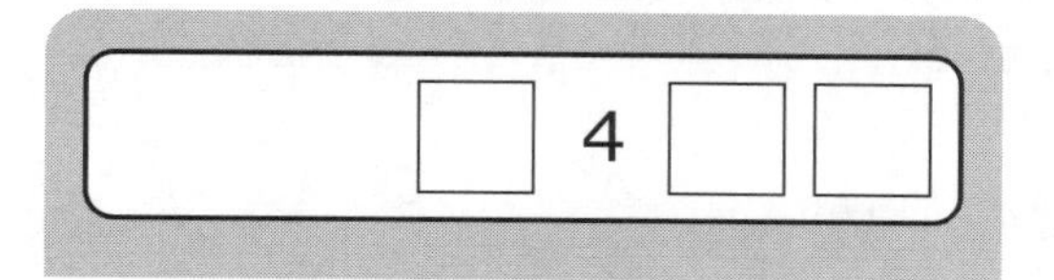

Write **all** the digits that might be on the calculator display.

2 marks

Level 5

Stamps are **19p** each.

Gwyn wants to buy **9** stamps.

He knows that he will have to pay **less than £2**.

a Show how you can tell that he will have to pay less than £2 **without** working out the exact answer. *1 mark*

b Gwyn buys **9** stamps at **19p** each.

Work out exactly how much he must pay. *1 mark*

A4.1HW Finding factors, multiples and primes

Learn these divisibility tests.

Divisible by	Rule
2	the last digit ends in 0, 2, 4, 6, 8
3	the sum of the digits is divisible by 3
4	the last two digits are divisible by 4
5	the last digit is 0 or 5
6	it is divisible by 2 and 3
8	half of it is divisible by 4
9	the sum of the digits is divisible by 9

1 Use the divisibility tests to work out and explain which of these two-digit numbers are prime.

a 97 **b** 79 **c** 51 **d** 91 **e** 63 **f** 73

2 To work out if a number is divisible by a two-digit number, you can apply two of the divisibility tests above.

For example:

Divisible by	Rule
12	the number is divisible by 3 and 4, the sum of the digits is divisible by 3 **and** the last two digits are divisible by 4

Hint: 3 and 4 are factors of 12.

Write a rule to test if a number is divisible by:

a 18 **b** 20 **c** 48

3 Which of these numbers are multiples of 12, 15, 18, 20 or 48?

a 180 **b** 210 **c** 216 **d** 1040

4 Find the highest common factor (HCF) of:

a 9 and 12 **b** 24 and 36

c 18 and 30 **d** 48 and 72

Hint: List the factors of each number.

5 Find the lowest common multiple (LCM) of:

a 4 and 6 **b** 8 and 7

c 12 and 15 **d** 16 and 36

Hint: List the multiples of each number.

A4.2HW Using prime factors

1 By writing each number as a product of its prime factors, find the HCF and LCM of the following pairs of numbers.
The first one is done for you.

a 12 and 18

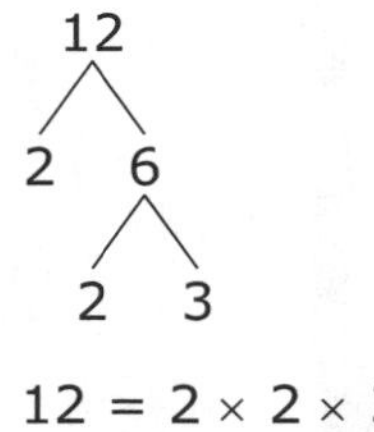

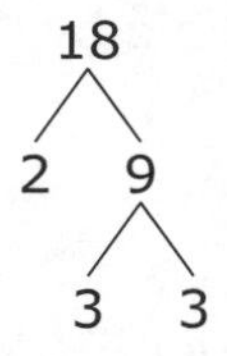

$12 = 2 \times 2 \times 3$
$= 2^2 \times 3$

$18 = 2 \times 3 \times 3$
$= 2 \times 3^2$

Write the factors in a Venn diagram:

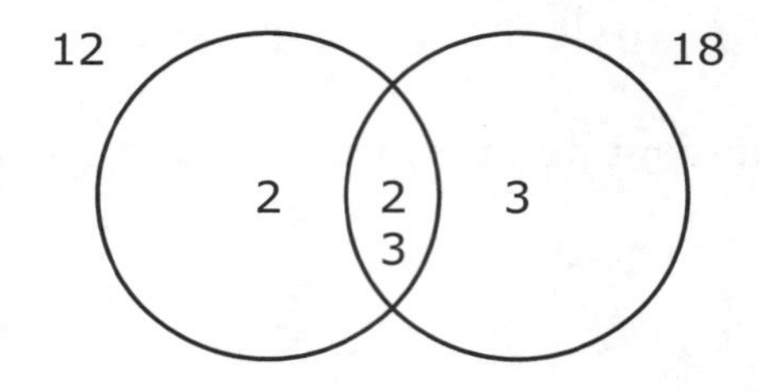

$\text{HCF} = 2 \times 3 = 6$

$\text{LCM} = 2 \times 2 \times 3 \times 3 = 36$

b 36 and 54

c 26 and 64

d 65 and 126

e 75 and 164

f 80 and 124

2 **Challenge**

Explain why the method works for finding HCF and LCM.

A4.3HW Using index notation

Remember:

$7^2 = 49$ and $\sqrt{49} = 7$

$7^3 = 343$ and $\sqrt[3]{343} = 7$

1 Work out the value of each power or root contained in the box.

2 Write your answers, with their letters, in order starting with the smallest.

3 What do the letters spell?

A $\sqrt{100}$	**C** 3^3	**E** 8^2
R $\sqrt{25}$	**A** $\sqrt[3]{64}$	**D** 5^2
Q $\sqrt{4}$	**B** 7^2	**S** 5^3
E 2^3	**S** 1^2	**U** $\sqrt{9}$
S $\sqrt{81}$	**N** 4^2	**U** 6^2

A4.4HW Solving problems involving indices

1 Copy and complete this table of powers of 2:

Power of 2	Value
2^0	
2^1	2
2^2	
2^3	8
2^4	
2^5	
2^6	

You can make 5 by adding powers of 2:

$$2^0 + 2^2$$
$$= 1 + 4$$
$$= 5$$

You can make 41 by adding powers of 2:

$$2^0 + 2^3 + 2^5$$
$$= 1 + 8 + 32$$
$$= 41$$

2 Write addition sums using powers of 2 for these numbers:

a 17 **b** 22 **c** 100 **d** 50

e 46 **f** 6 **g** 37 **h** 81

Hint: Use your table from question **1** to help you.

3 Write five calculations of your own, using powers of 2.

What is the largest number you can make?

A4.5HW Plotting linear graphs

1 **a** Copy and complete the table of values for the function $y = 2x + 3$.

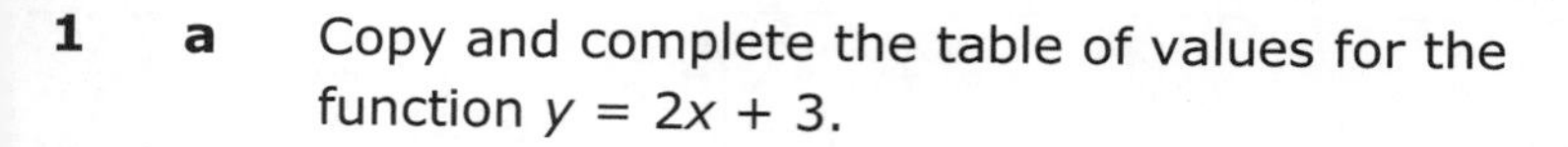

x	$^-3$	$^-2$	$^-1$	0	1	2	3
y	$^-3$			3			

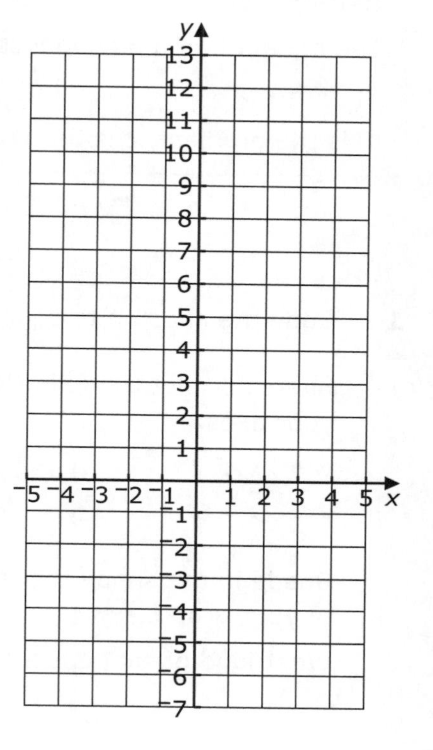

b Write down the coordinate pairs.

c Copy the grid and plot the coordinate pairs. Join the coordinate pairs and extend to the edges of the grid.

d The following coordinates lie on the straight line $y = 2x + 3$.

$(x, 6)$ $(^-4, y)$ $(x, 0)$

Use the graph to complete the coordinates.

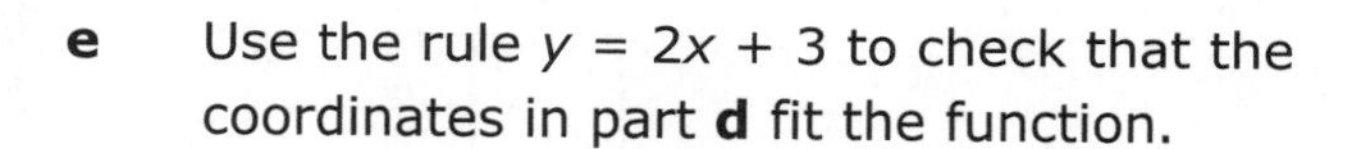

e Use the rule $y = 2x + 3$ to check that the coordinates in part **d** fit the function.

2 **a** Draw the graphs of $y = x + 2$ and $y = 2x - 3$ for x values $^-3$ to 3 on a suitable size grid.

> **Hint:** Look for the smallest and largest y values.

b Write down the coordinates of the point where the two lines cross.

Check that these coordinates fit both functions.

A4.6HW Straight-line graphs

Remember:

The equation of a diagonal straight line is

$y = mx + c$.

m is the gradient, c is the y-intercept.

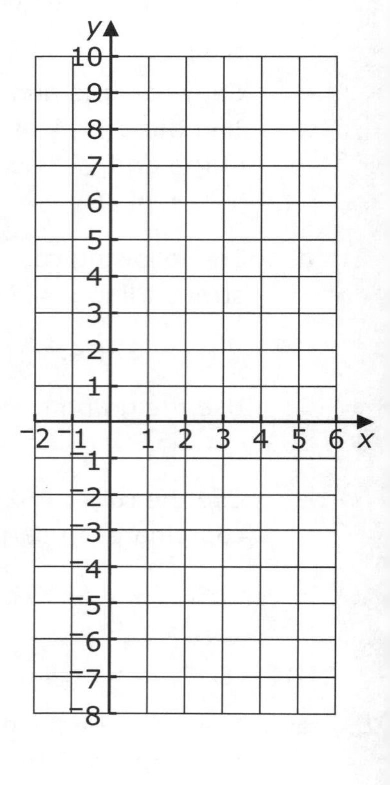

1 Copy the axes on to squared paper.

Sketch or plot the following straight-line graphs on your axes.

$y = 2x + 2$	$y = 4$	$y = {}^{-}2x + 10$
$y = 2x - 8$	$y = {}^{-}2$	$y = {}^{-}2x$

Shade in the shape that the lines enclose in the centre of your graph.

What kind of shape is it?

2 Match each function to its graph.

a $y = 3x - 1$ **b** $y = 2x$ **c** $y = x + 2$

Graph 1

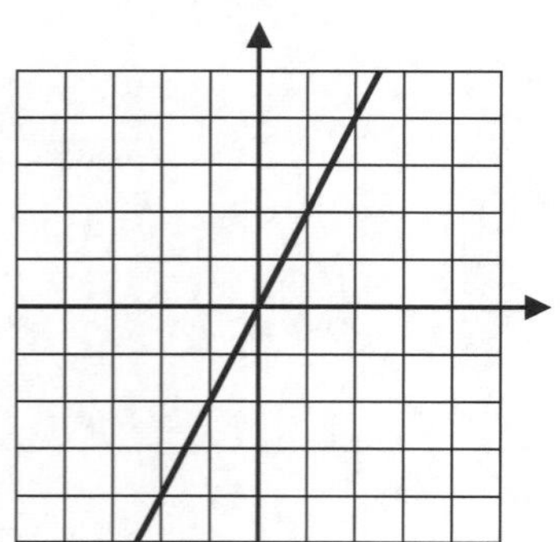

Graph 2

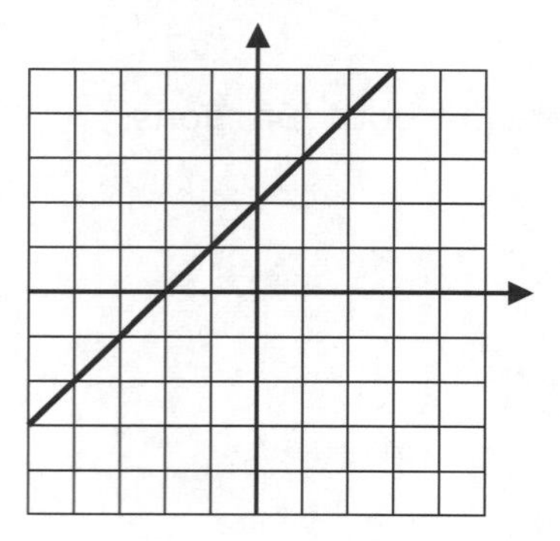

Graph 3

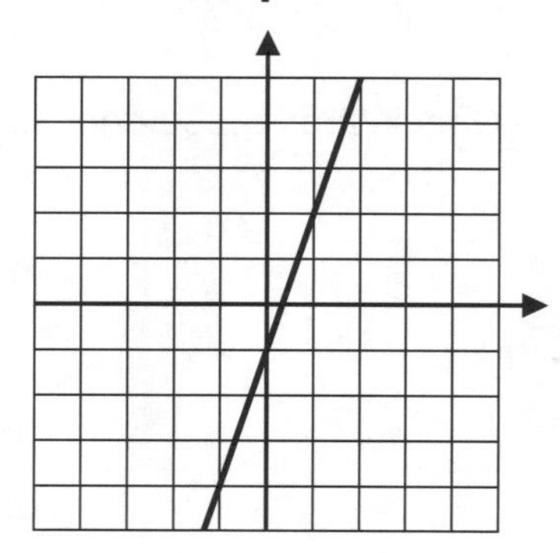

A4.7HW Recognising straight-line graphs

Dominoes

Make a set of 16 dominoes so that when they are played they make a complete loop.

The loop has been started for you.

You decide which graph goes here!	$y = 2x$
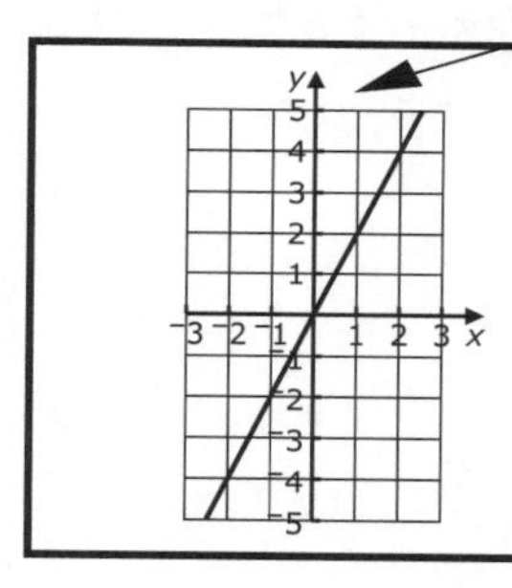	$y = x + 2$
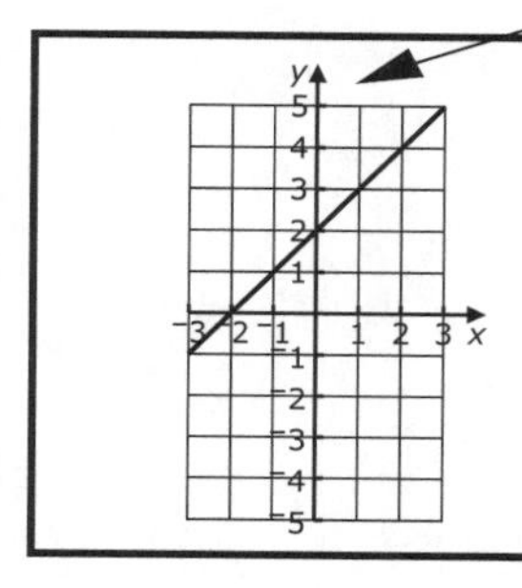	$y = 3x - 2$
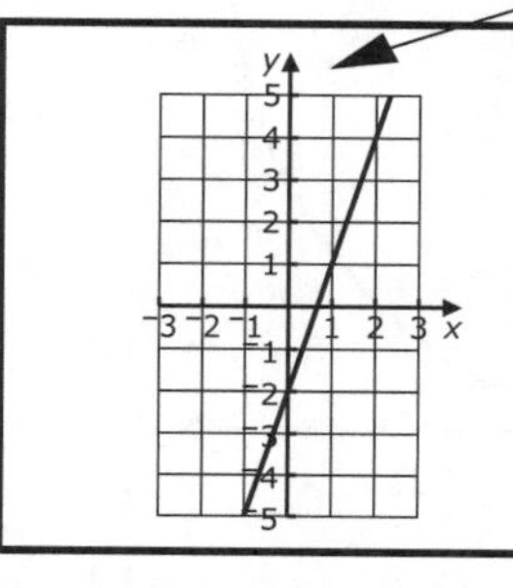	You decide which equation goes here!

Play a game of dominoes with a partner.

A4.8HW Using graphs

1 Give an explanation for the shape of the following graphs from A to F. Each explanation has been started for you.

a Temperature of bath water

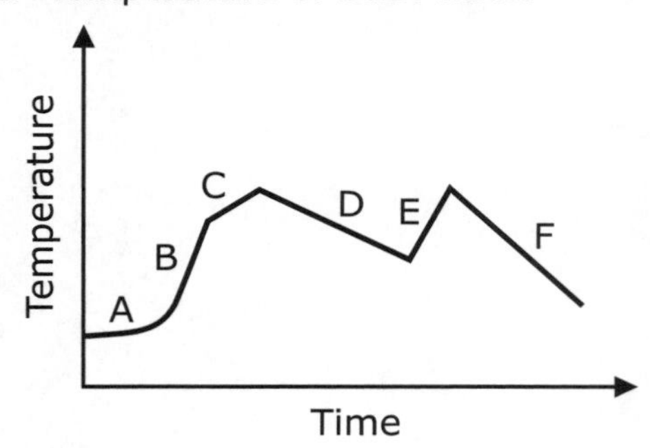

A: Hot water tap is running but is not hot yet.

B: Hot water tap is running and it's hot.

C:

b Weight of bag of sweets

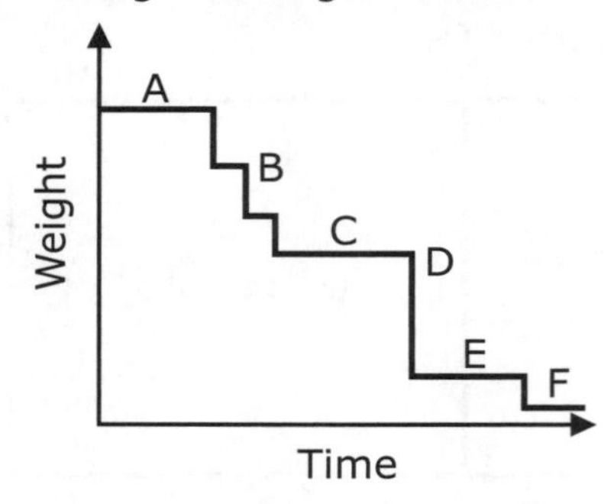

A: No sweets are taken from the bag.

B: A few sweets are taken out.

C:

2 The container below is being steadily filled with water.

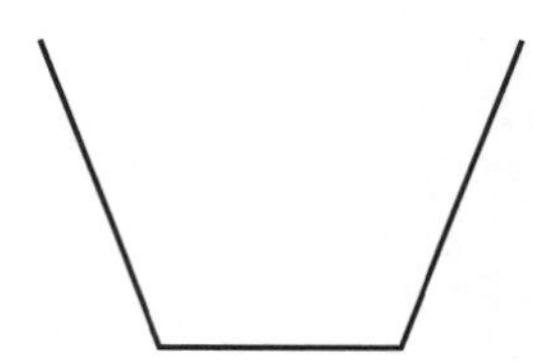

Which of the graphs shows the relationship between time and height of water in the container?

a

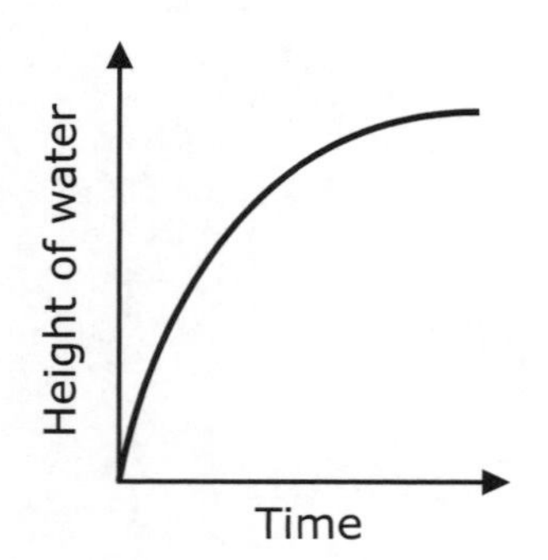

b

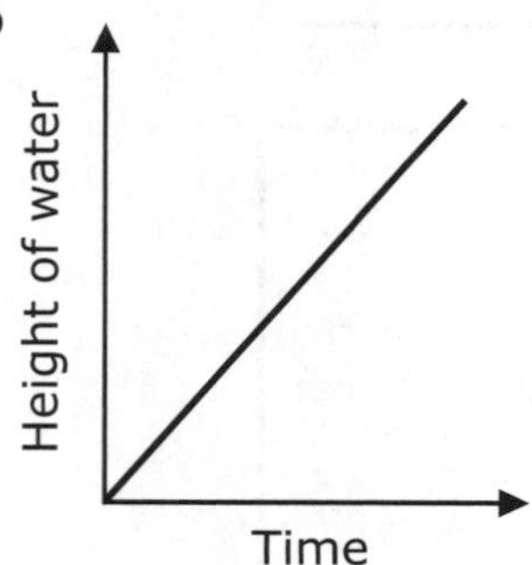

c

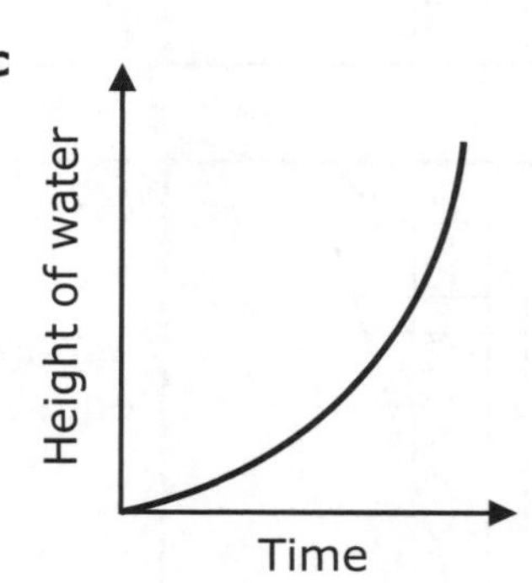

Explain your choice.

A4.9HW Interpreting distance–time graphs

1 This is a distance–time graph for Elaine and her friend's trip to the shops.

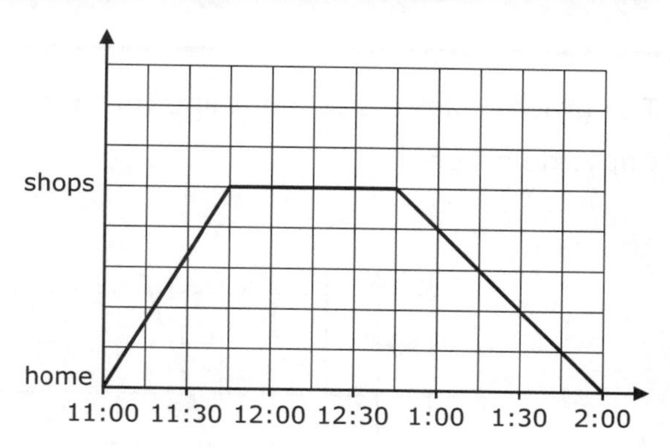

a How long did it take them to get to the shops?

b How long were they at the shops?

c Was the journey back from the shops faster or slower than the journey there?

d How long were they travelling for?

e How long were they away from home in total?

2 This graph shows Miss Millward's journey to work.

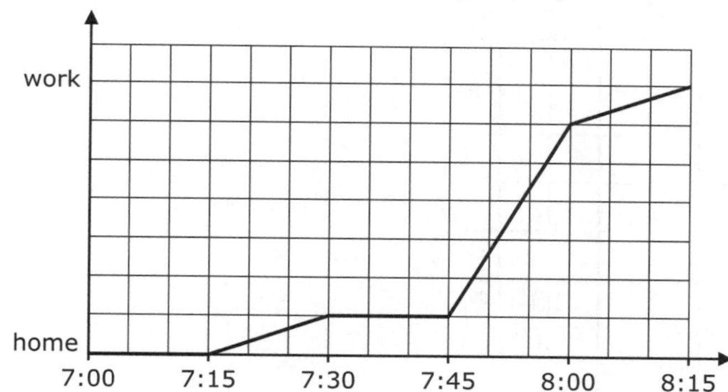

Describe her journey in words.

3 **a** Draw a distance–time graph, on a copy of the grid opposite, for the following train journey:

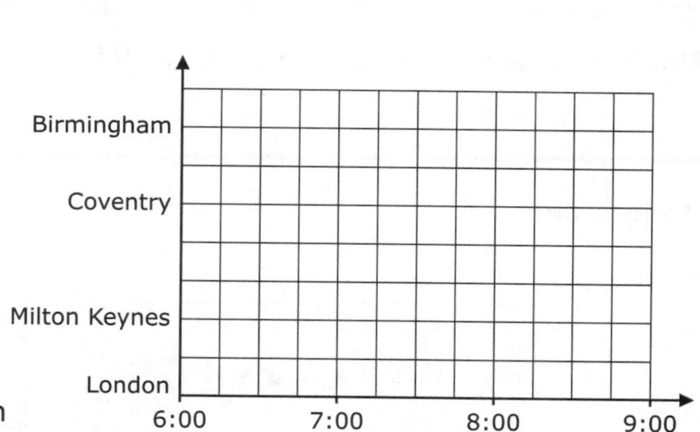

A: Train leaves London at 6 pm

B: Train stops at Milton Keynes at 6.30 pm

C: Train leaves Milton Keynes at 6.45 pm

D: Train stops at Coventry at 7.15 pm

E: Train leaves Coventry at 7.45 pm

F: Train arrives in Birmingham at 8.15 pm

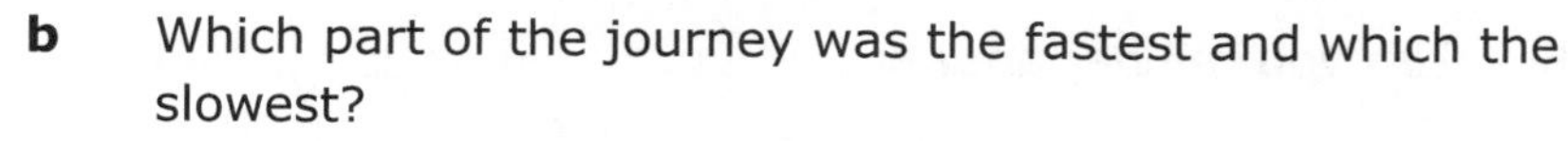
b Which part of the journey was the fastest and which the slowest?

c For how long was the train stationary?

Level 4

The graphs shows a straight line through the point (2, 5).

Copy the graph.

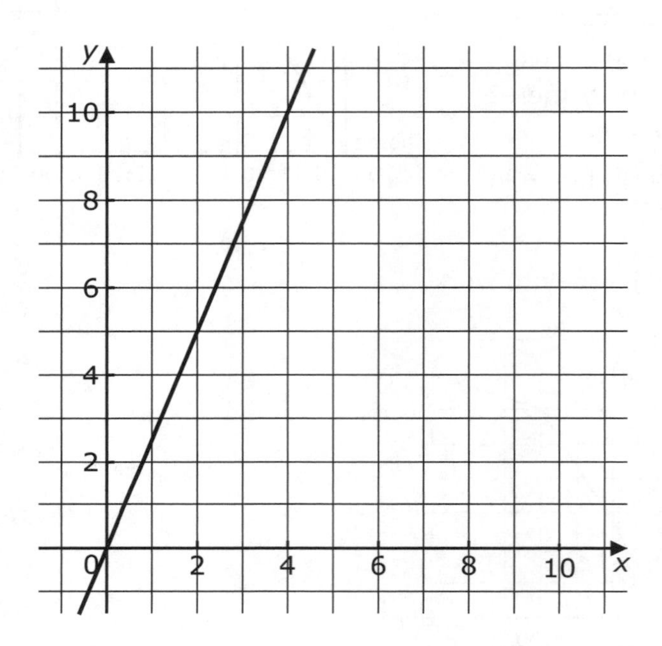

Use a ruler to draw another straight line that is **parallel** to this line.

The line **must** go through the point **(3, 0)**. *1 mark*

Level 5

Marie wrote:

$594 \times 25 = 14\,852$

Without doing the calculation, how can you tell her answer is **wrong**? *1 mark*

D2.1HW Revising probability

1 Draw and label a probability scale. Use these labels.

Impossible

0

Evens chance

1

2 Amit and Denise each buy an Indian snack pack with the contents shown.

Amit picks a snack at random from his pack.

Find the probability that he picks:

a a bhaji

b a samosa

c a bhaji or a pakora.

Denise eats three of the pakoras from her pack.
She then picks another snack at random.

d Find the probability that she picks a samosa.

3 **a** What is the probability of rolling a six on an ordinary dice?

b Stephen rolls a dice 48 times and gets 12 sixes.
What is the probability of rolling a six on Stephen's dice?

c Do you think Stephen's dice is fair? Explain your answer.

D2.2HW The probability of an event not occurring

Remember:

- If you know that the probability of an event occurring is p, then the probability of the event not occurring is $1 - p$.

1 The probability of Sushma completing her homework in 15 minutes is 0.45.

Work out the probability that she does not complete her homework in 15 minutes.

2 The table shows the probability, p, of some events.
Copy the table and complete it to show the probability that each event does **not** occur, $1 - p$.

p	0.36	25%	$\frac{3}{8}$	0.72	63%	$\frac{2}{9}$
$1 - p$						

3 Here are 10 cards:

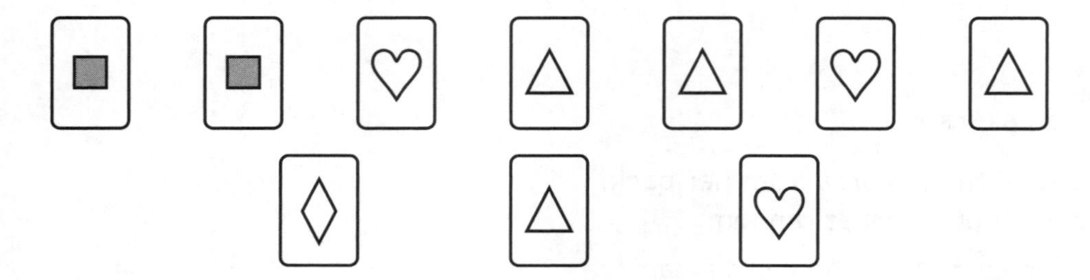

The cards are shuffled and spread face down.

One card is picked at random.

Give your answers to the following questions as percentages.

a What is the probability that the card picked is a triangle?

b What is the probability that the card picked is a heart?

c What is the probability that the card picked is **not** a triangle?

d What is the probability that the card picked is **not** a square?

D2.3HW Sample space diagrams

1 A card is picked from a pack of playing cards and a coin is tossed.

The possible outcomes for the coin are:

Heads

Tails

The possible outcomes for the cards are:

Spades

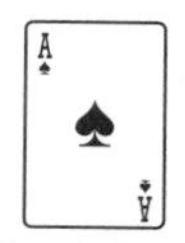
Clubs

Hearts

Diamonds

a Copy and complete this sample space diagram to show the possible outcomes.

	Card suit			
	Spades	Hearts (He)	Clubs	Diamonds
Heads	(H, S)	(H, He)		
Tails				

Use your sample space diagram to work out the probability that:

b the coin will show Tails

c the card will show Clubs

d the coin will show Heads and the card will show a Diamond.

2 These two fair spinners are spun together.
The scores on the spinners are added together.

a Draw a sample space diagram to show the possible outcomes.

b Use your diagram to find the probability that the total will be an odd number.

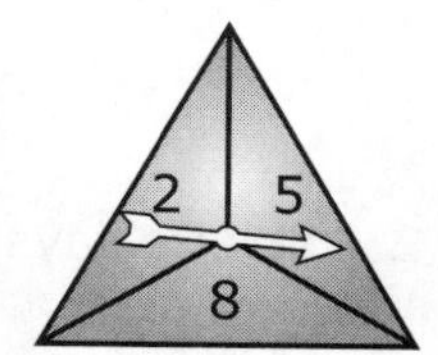

Hint: The square spinner has four outcomes

3 Caitlin and Sion each have three number cards.

They each pick **one** of their cards at random, and they add the numbers on the two cards.

a Draw a sample space diagram to show the possible totals.

Use your sample space diagram to work out the probability that the total is:

b exactly 3

c less than 5

d an even number

e more than 8.

D2.4HW Experimental probability

1 The cricket teams in a league collect the results for their first 50 matches in a season.

Home win	**35**
Away win	**15**

Use these results to estimate the probability in future cricket matches of:

a a home win

b an away win.

For the next 50 matches the results were

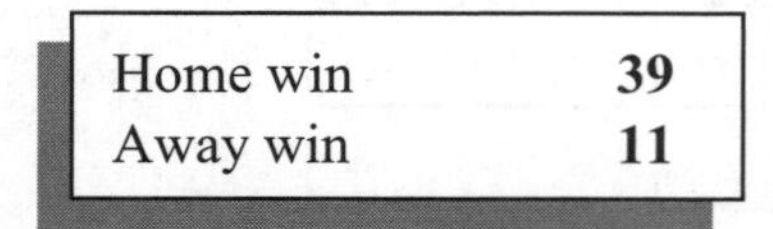

Home win	**39**
Away win	**11**

Combine both sets of results to estimate the probability in future cricket matches of:

c a home win

d an away win.

2 Claire says:

'When I drop a piece of toast, the probability of it landing butter side down is 75%'.

Describe an experiment you could do to test whether Claire's estimate of the probability is correct.

D2 Probability

SAT QUESTIONS

Level 4

A class has some gold tokens and some silver tokens.

The tokens are all the same size.

a The teacher puts **4 gold tokens** and **1 silver token** in a bag.

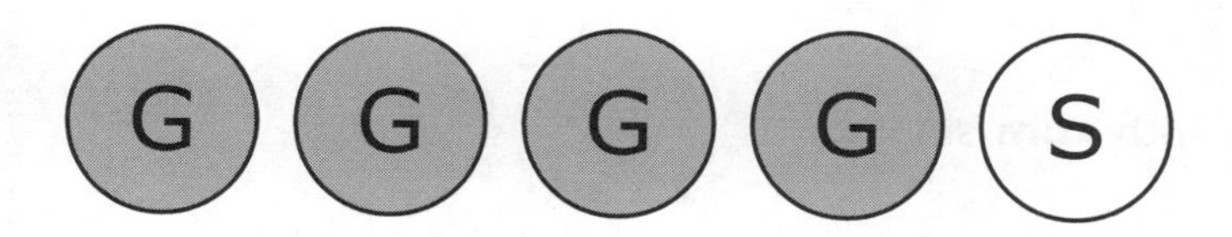

Leah is going to take one token out of the bag without looking.

She says:

> There are two colours, so it is **just as likely** that I will get a gold token as a silver token.

Explain why Leah is **wrong**. *1 mark*

b How many **more silver** tokens should the teacher put in the bag to make it just as likely that Leah will get a gold token as a silver token? *1 mark*

c Jack has a different bag with **8** tokens in it.

It is **more likely** that Jack will take a gold token than a silver token from his bag.

How many **gold** tokens might there be in Jack's bag? *1 mark*

Level 5

Les, Tom, Nia and Ann are in a singing competition.

To decide the order in which they will sing all four names are put into a bag.

Each name is taken out of the bag, one at a time, without looking.

a Write down **all** the possible orders with **Tom** singing **second**. *2 marks*

b In a different competition there are 8 singers.

The probability that Tom sings second is $\frac{1}{8}$.

Work out the probability that Tom does **not** sing second. *1 mark*

S3.1HW Bearings

This map shows mountains in the Lake District.

Scafell Pike is the highest mountain in England.

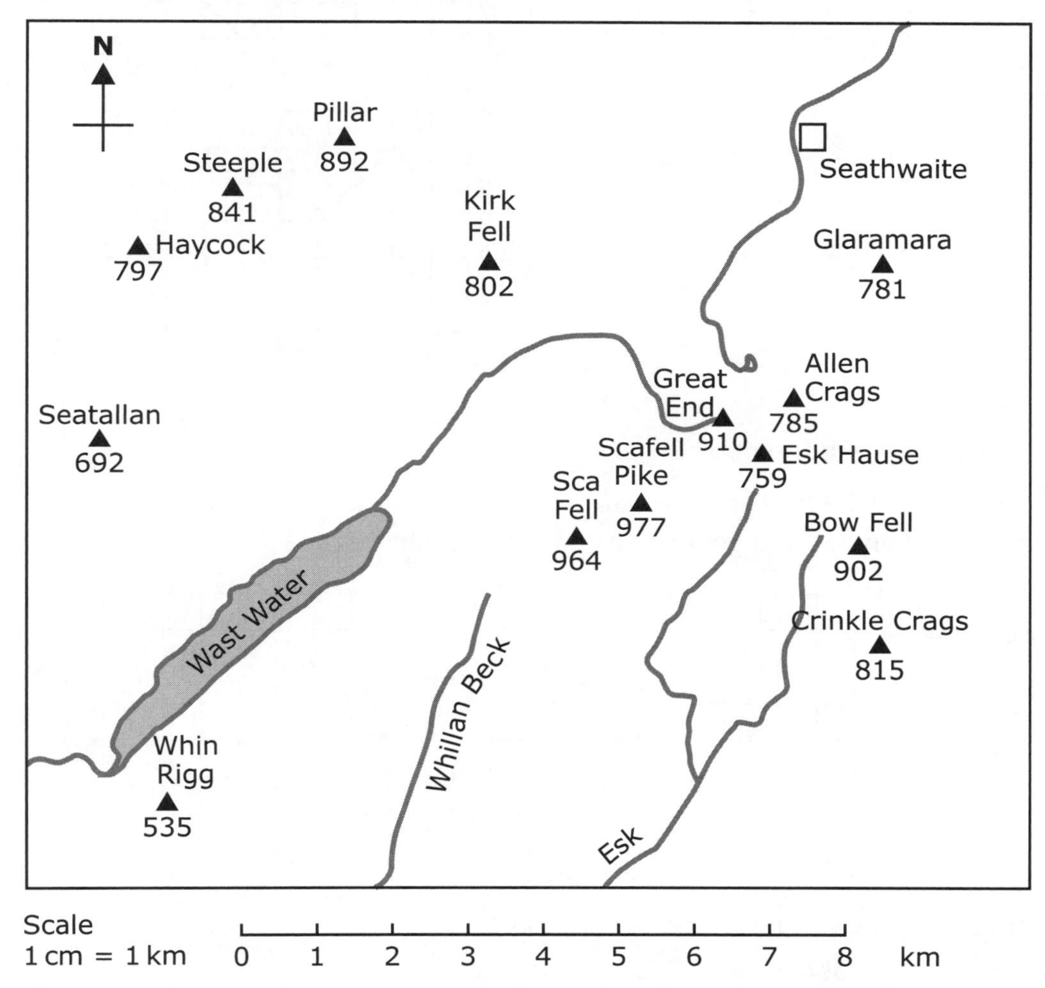

1 Write down the names of the mountains that are on these bearings from Scafell Pike.

a 043º **b** 122º **c** 238º **d** 277º **e** 308º

2 Write down the bearings of these mountains from Scafell Pike.

a Great End **b** Bow Fell **c** Whin Rigg **d** Haycock **e** Pillar

3 Estimate the distance, in km, from Scafell Pike to:

a Sca Fell **b** Bow Fell **c** Glaramara

d Kirk Fell **e** Seathwaite

4 It is impossible to see Whin Rigg from Scafell Pike, even on a fine day. Why?

S3.2HW Describing transformations

1 Write down the translation that moves the shaded face to:

a A

b B

c C

d D

e E

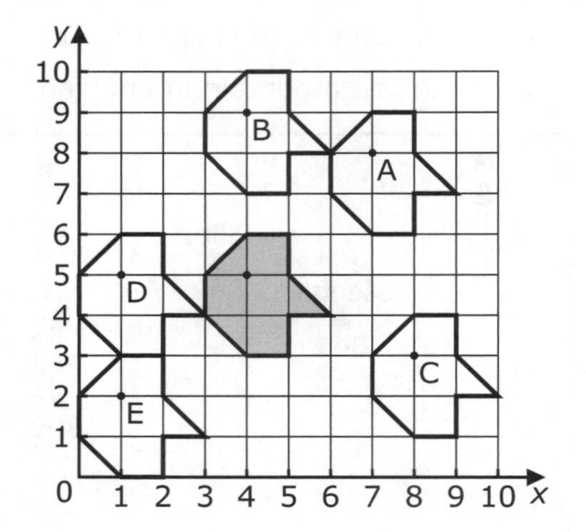

2 Reflect each shape in the mirror line.

Name the new shape formed.

Mark the equal angles and equal sides on the new shape.

a

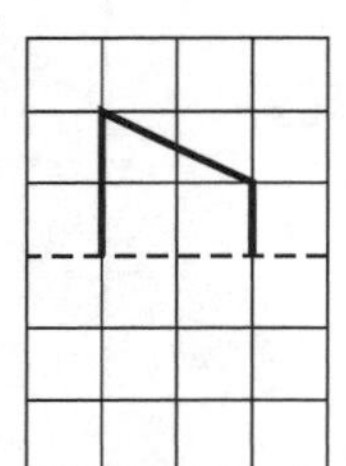

b

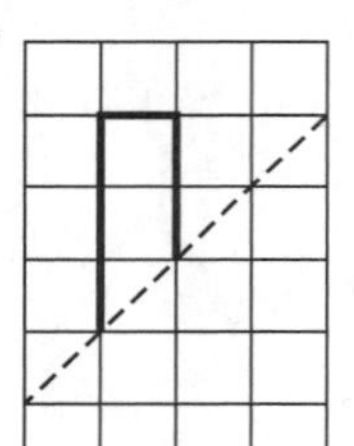

c

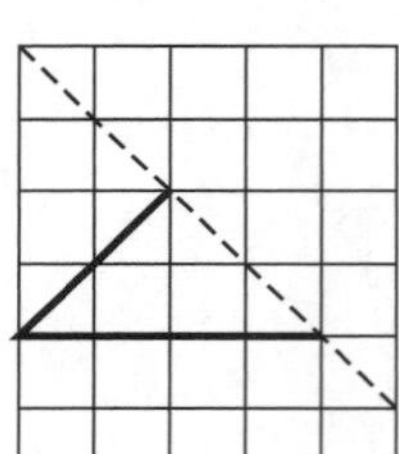

d

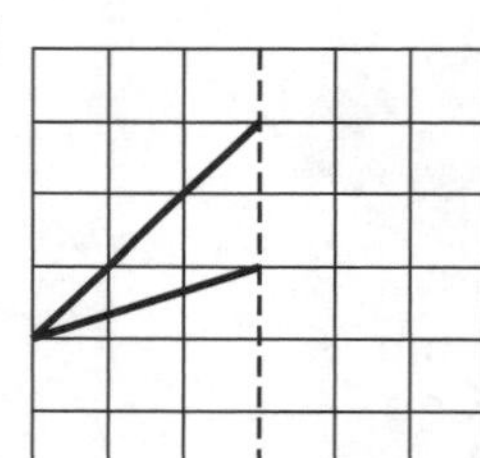

e

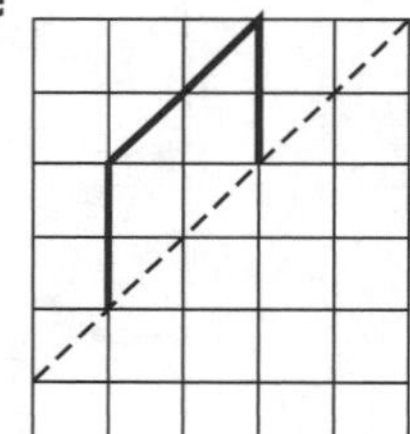

3 Copy these shapes on to squared paper.

Rotate each shape about the dot (•) through 180°.

Name the new shape formed by adding the shape and rotation together.

Mark the equal angles and equal sides on the new shape.

a

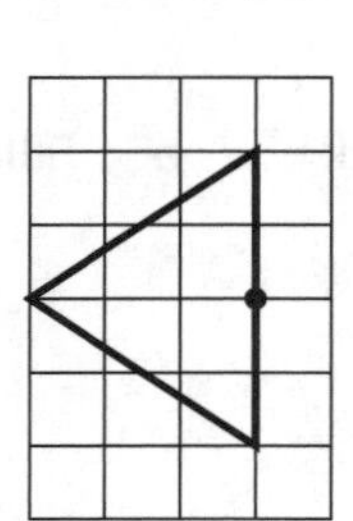

b

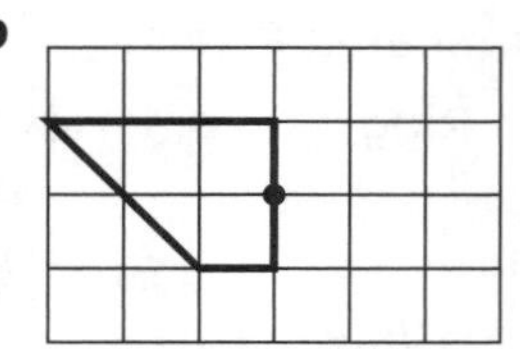

c

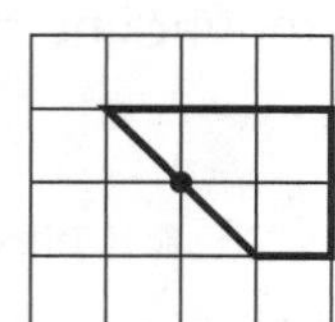

d

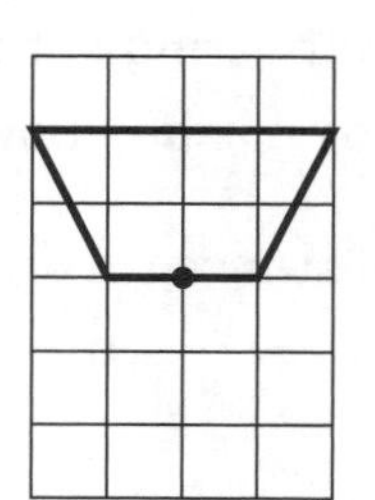

e

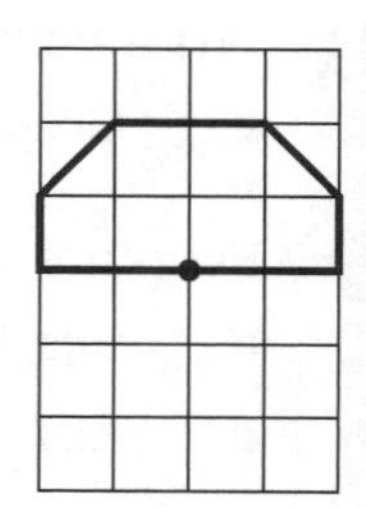

S3.3HW Combining transformations

1 Copy the diagram on to squared paper.

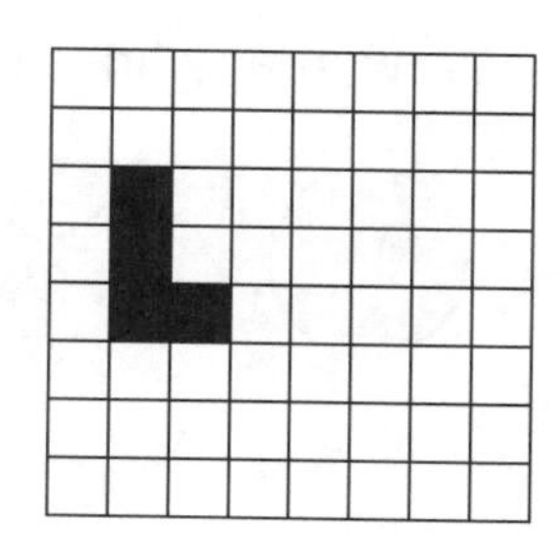

a Translate the black shape by $\begin{pmatrix} \text{2 right} \\ \text{2 up} \end{pmatrix}$.
Label the new shape A.

b Translate the shape A by $\begin{pmatrix} \text{3 right} \\ \text{1 down} \end{pmatrix}$.
Label the new shape B.

c What single transformation moves the black shape directly to the shape B?

2 **a** Copy this arrowhead on to squared paper and cut it out.

b Use repeated 180° rotations about the midpoints of the sides to tessellate the arrowhead.

c Repeat this process for a parallelogram.

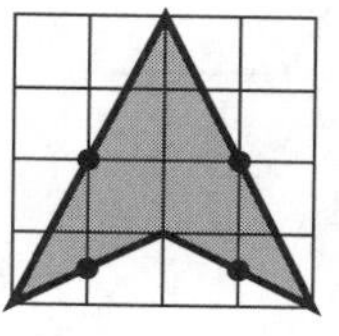

3 Copy this diagram on to squared paper.

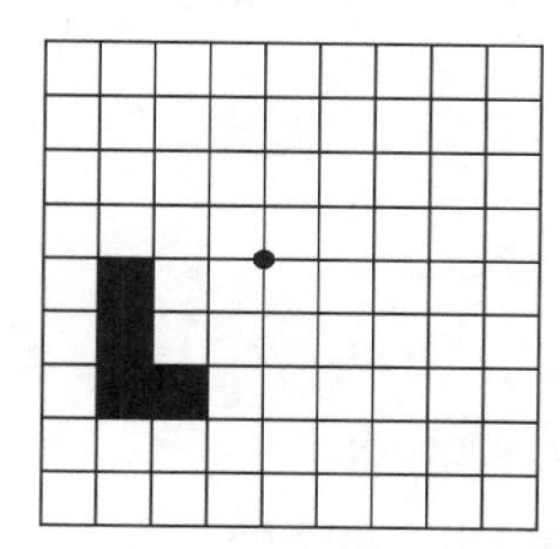

a Rotate the black shape through 90° clockwise about the dot.
Label the new shape A.

b Rotate the shape A through 90° clockwise about the dot.
Label the new shape B.

c What single transformation moves the black shape directly to the shape B?

Symmetries of polygons

1 **a** Copy each shape.

b Draw in any lines of symmetry.

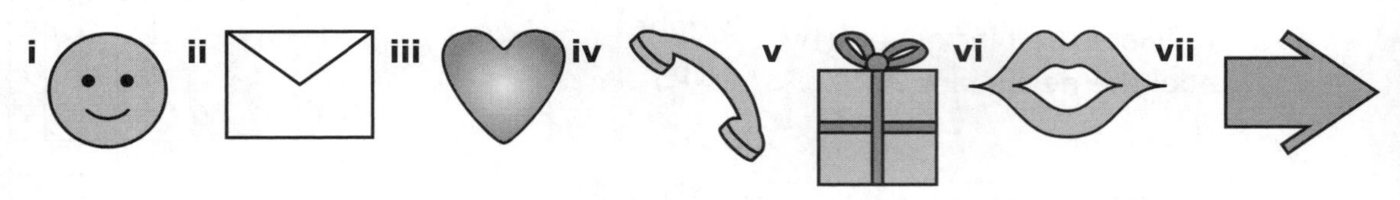

2 State the order of rotational symmetry for each shape.

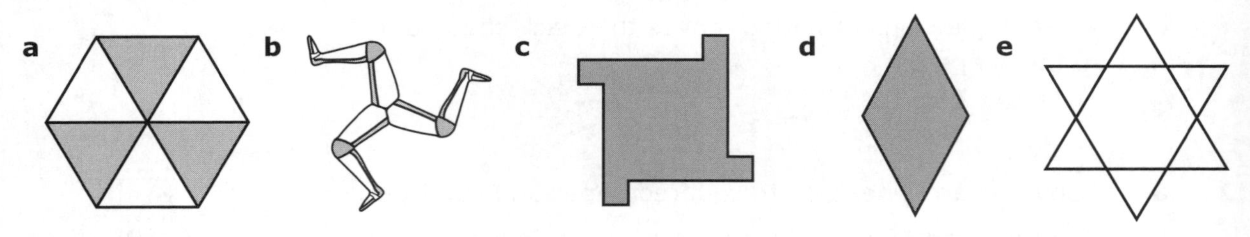

3 **a** How many lines of symmetry has a regular decagon?

b What is the order of rotational symmetry for a regular decagon?

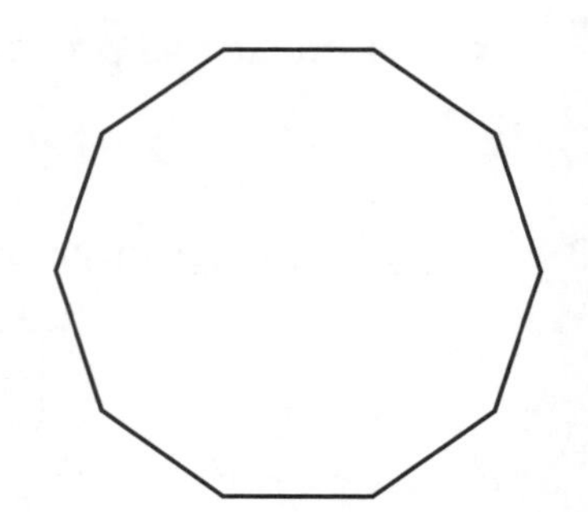

4 An equilateral triangle has rotational symmetry of order 3.

What geometrical properties does this mean the triangle has?

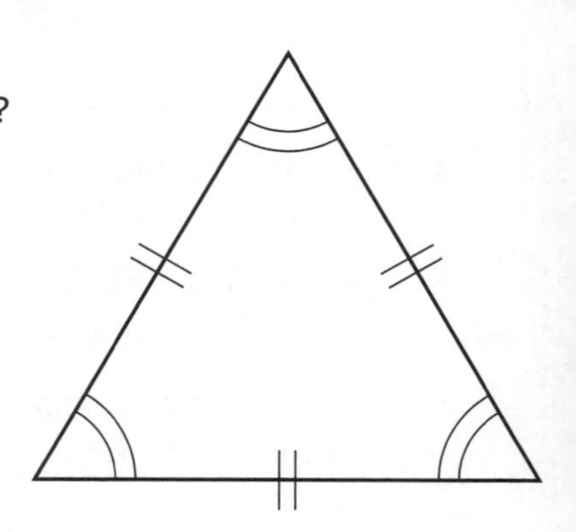

S3.5HW Enlargements

1 Find the scale factor of these enlargements.

a

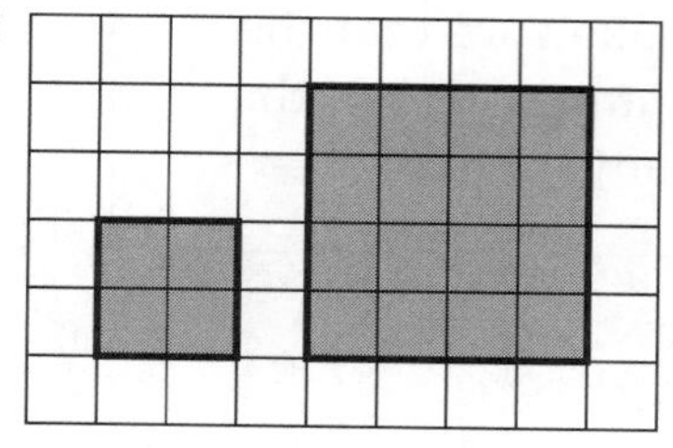

b

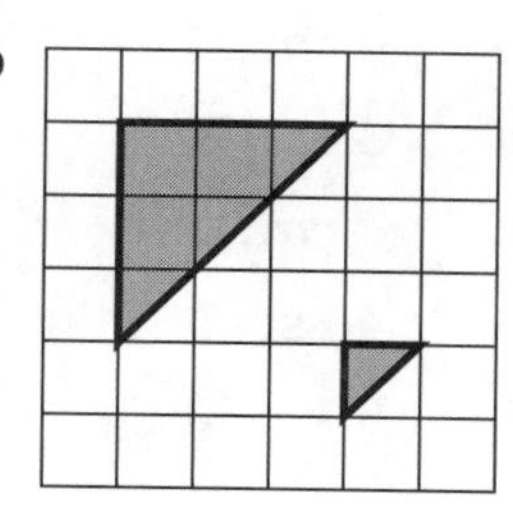

c

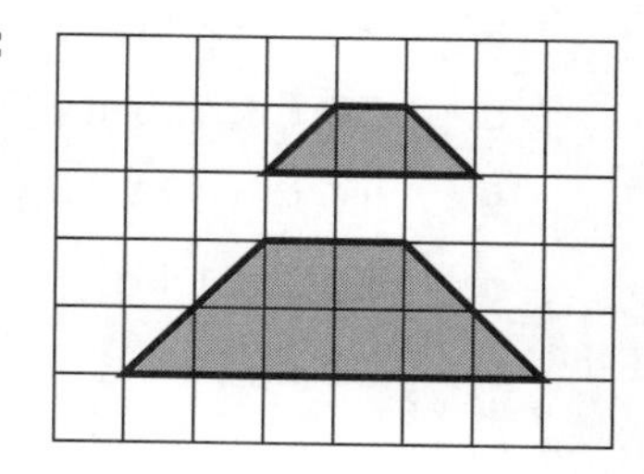

2 Copy these shapes on to squared paper.

Draw the enlarged shape.

a

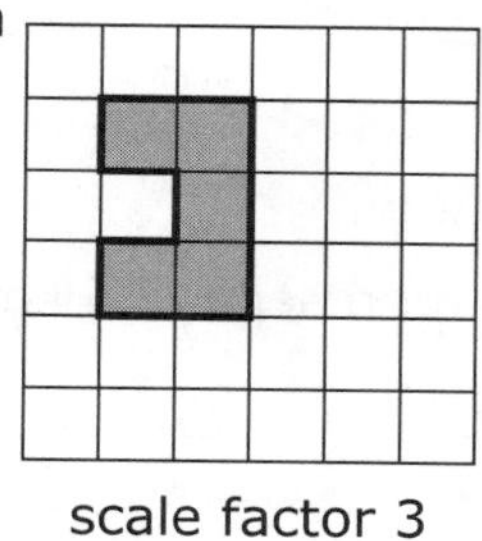

scale factor 3

b

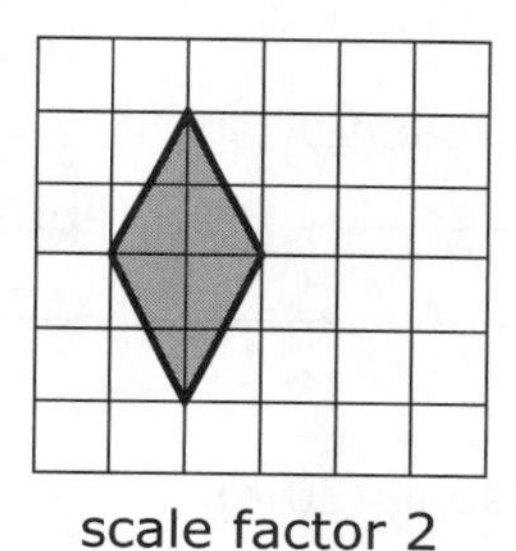

scale factor 2

c

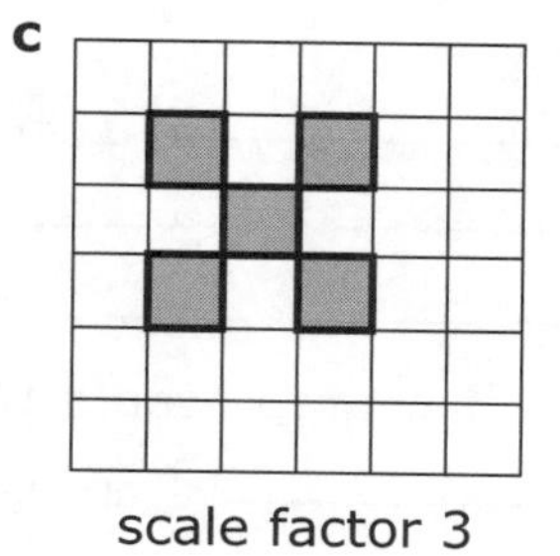

scale factor 3

d

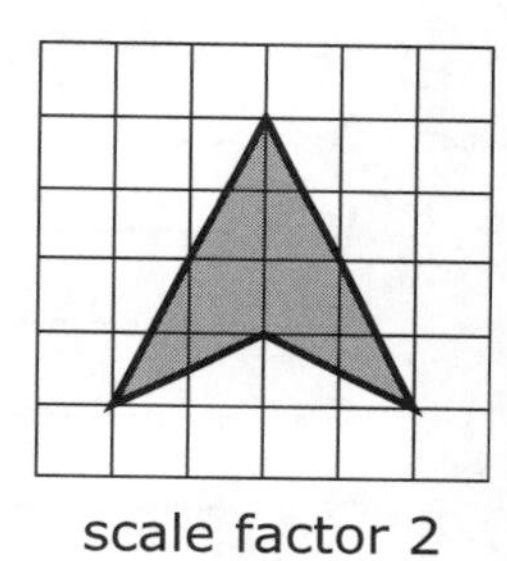

scale factor 2

3 The shaded shapes are enlarged to give shapes A, B and C.

a

C

B

A

b

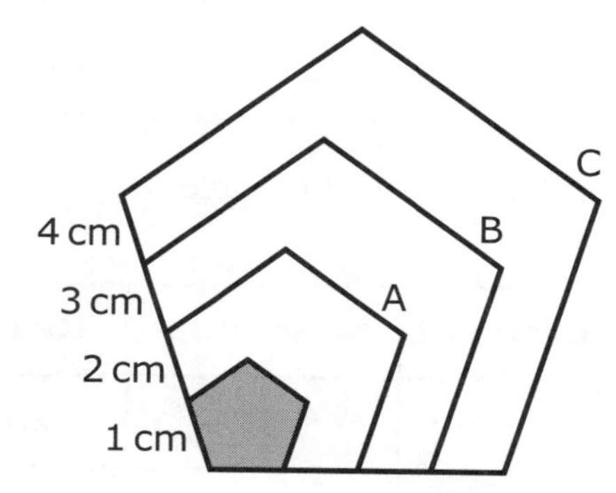

Copy and complete these tables.

Shape	Scale factor	Perimeter
Shaded square	1	4 cm
Square A		
Square B		
Square C		

Shape	Scale factor	Perimeter
Shaded pentagon	1	5 cm
Pentagon A		
Pentagon B		
Pentagon C		

S3.6HW Scale drawings

1 Express these ratios in their simplest form:

a 4 kg:12 kg **b** 9 cm:12 cm

c 30 ml:45 ml **d** 8 m:12 m

e 16 mm:24 mm **f** 4 cm:1 m

g 500 g:1 kg **h** 40 sec:1 min

i 30 cl:1 l **j** 5 kg:500 g

Hint:
Make sure that the numbers are in the same units.

2 Measure the length of these lines.

Use the scale to calculate the actual distance that each line represents.

a ______________________________ 1 cm represents 1 m

b ____ 1 cm represents 2 km

c ____ 1 cm represents 2.5 km

d _______________ 1 cm represents 10 m

e ___________________________ 1 cm represents 20 m

3 Convert the scales in question **2** into ratios.

4 A model boat is built using a scale of 1:20.

Copy and complete this table.

Measurement	Model boat	Real boat
Width		250 cm
Length		900 cm
Number of portholes		3

5 Make a scale drawing of your bedroom, including the furniture.
Choose a suitable scale.

Level 4

I have a square grid and two rectangles.

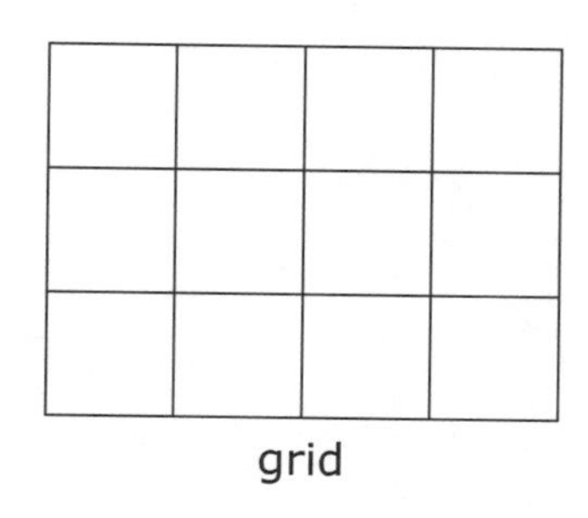
grid

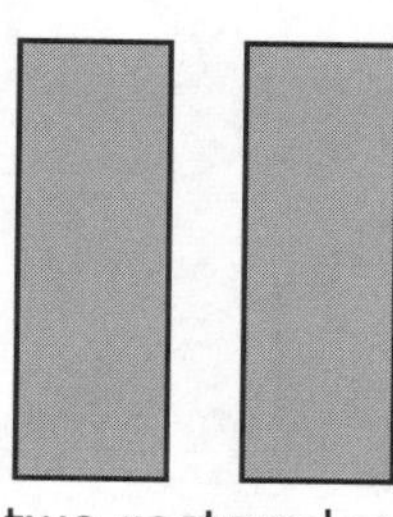
two rectangles

I make a pattern with the grid and the two rectangles.

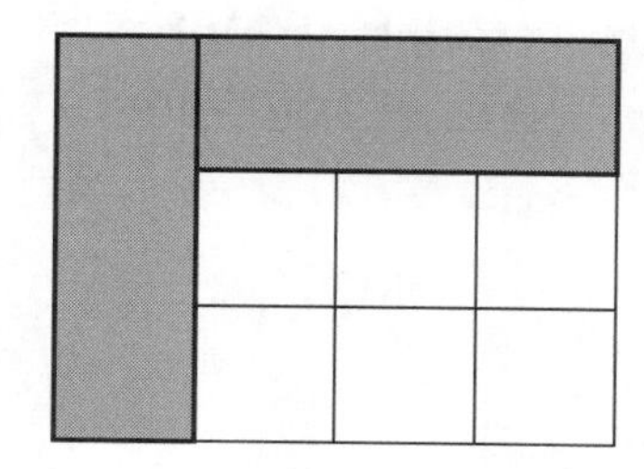

The pattern has **no** lines of symmetry.

a Copy the grid.

Put both rectangles on the grid to make a pattern with **only one** line of symmetry.

You must **shade** the rectangles. *1 mark*

b Put both rectangles on the grid to make a pattern with **rotation** symmetry of **order 2**.

You must **shade** the rectangles. *1 mark*

Level 5

This cuboid is made from **4** small cubes.

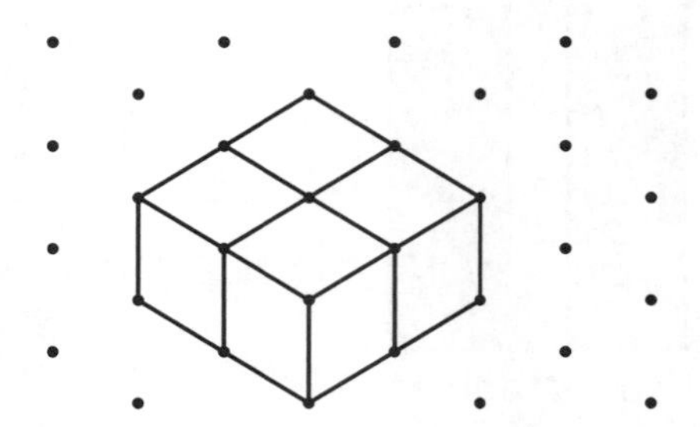

a On isometric dotty paper, draw a cuboid which is **twice** as **high**, **twice** as **long** and **twice** as **wide**. *2 marks*

b Graham made this cuboid from **3** small cubes.

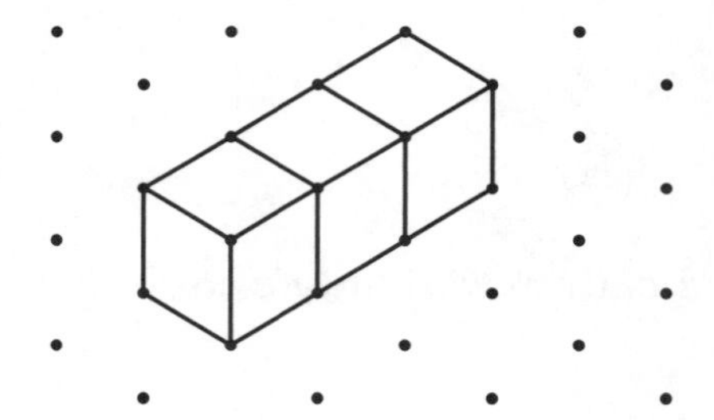

Mohinder wants to make a cuboid which is **twice** as **high**, **twice** as **long** and **twice** as **wide** as Graham's cuboid.

How many small cubes will Mohinder need altogether? *1 mark*

P1.1HW Making sense of the problem

Janine is tiling her bathroom using mosaic tiles.

She has 3 different sized tiles.

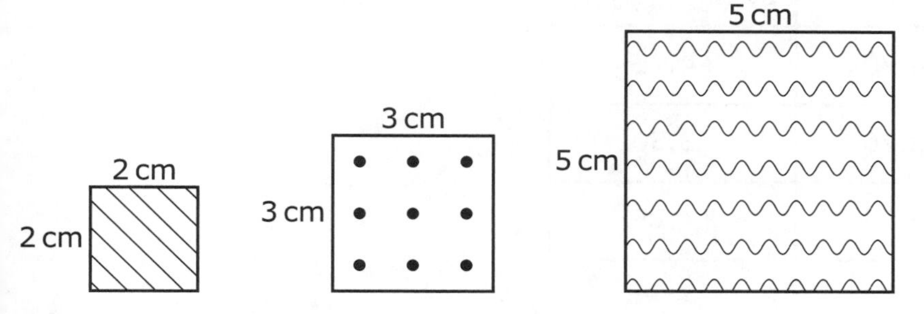

On the wall she has marked out an area 10 cm by 10 cm.

a How many of the 5 cm by 5 cm tiles will cover the area?

b How many of the 2 cm by 2 cm tiles will cover the area?

Janine puts three tiles on the wall like this:

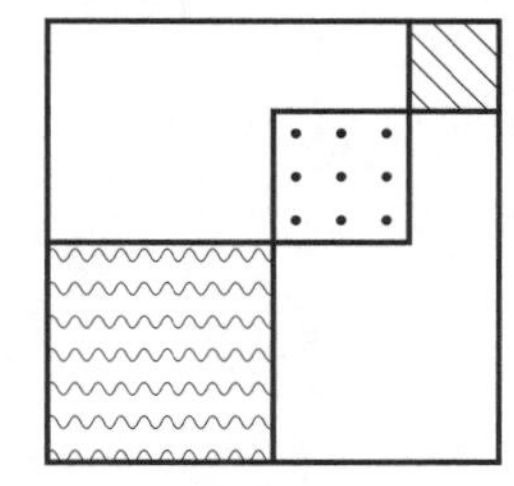

c Can Janine completely fill the 10 cm by 10 cm area without cutting any tiles?

Explain how you got your answer.

d Janine sees some 1 cm by 1 cm tiles in a catalogue.

What is the minimum number of 1 cm by 1 cm tiles she needs to order to fill her 10 cm by 10 cm area starting with the pattern of larger tiles shown above?

P1.2HW Reading tables and diagrams

1 This table shows a cinema's prices.

	Tuesday	All days except Tuesday
Adult	£4.20	£4.75
Child (16 or under)	£3.20	£3.50
Senior citizen (60 or over)	£3.75	£4.25

a Mrs Johnson (aged 32), her mother (aged 63) and her two children (aged 7 and 5) go to the cinema on Tuesday.
How much does it cost them?

b Sophie (17), Tom (15) and Cherry (16) go to the cinema on Wednesday.
How much does it cost them?

2 Samir did a survey in his school.
He asked students 'Do you like school dinners?'
This table shows his results.

	Yes	No	Don't know
Year 7	18	32	3
Year 8	28	21	5
Year 9	19	23	6

a How many students from Year 8 took part in the survey?

b In total, how many more students said 'No' than said 'Yes'?

P1.3HW Using a strategy

1 Sarah has £58.75 in her bank account.

Tracy has £37.15 less than Sarah.

How much does Sarah have in her account?

2 Jodie wants to buy a Hi-Fi system.

It costs £400.

There are two ways she can pay.

Pay now and get **15%** off!

Pay **£8.50** per week for 52 weeks

How much will she save if she pays now, with 15% off?

3 **a** 'Butcher's block' worktops cost £88.40 per square metre.
How much would it cost to cover this workspace?

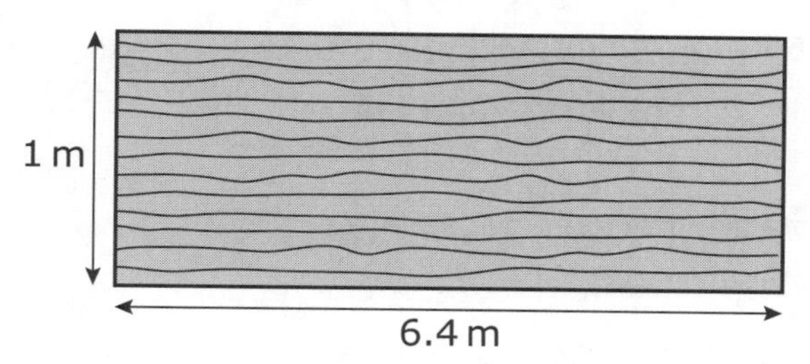

b Haricot beans cost £1.40 per kg.

How much would 32.6 kg of haricot beans cost?

c The ratio of the height of Paul to the height of Shoaib is 6 : 5.

If Shoaib is 1.6 m, how tall is Paul?

d Brown carpets cost £23.70 per square metre.

How much would it cost to carpet an area of 9.8 m^2 with brown carpet?

P1.4HW Proportional reasoning

> **Hint for questions 1 and 2:**
> First find the values for **one** gram.

1 The label on a tin of spaghetti hoops shows this information.

Average values per 100 g	
Energy	62 Cals
Protein	1.7 g
Carbohydrate	13 g
Fat	0.3 g

Work out how much energy, protein, carbohydrate and fat is provided by a 410 g tin of spaghetti hoops.

Give your answers to one decimal place whenever necessary.

2 The label on a jar of mayonnaise shows this information.

100 g typically provides	
Energy	721 Cals
Protein	11.3 g
Carbohydrate	13.2 g
Fat	63 g

A recipe for potato salad requires 65 g of mayonnaise.

Work out how much energy, protein, carbohydrate and fat is provide by 65 g of mayonnaise.

Give your answer to one decimal place wherever necessary.

3 Jim is on a special diet that means he should not have more than 4 g of salt in one meal.

He wants to have a pizza, ice cream and soft drink for dinner.

Pizza per 100 g	
Salt	1.5 g

Supersoft Ice cream per 100g	
Salt	2 g

Lemonade per 100 ml	
Salt	1.3 g

a Can he have 300 g pizza, 100 g ice cream and 150 ml lemonade?

b Suggest different portion sizes that Jim could have to fit his diet.

Geometrical reasoning

The diagram shows a shape ABC.

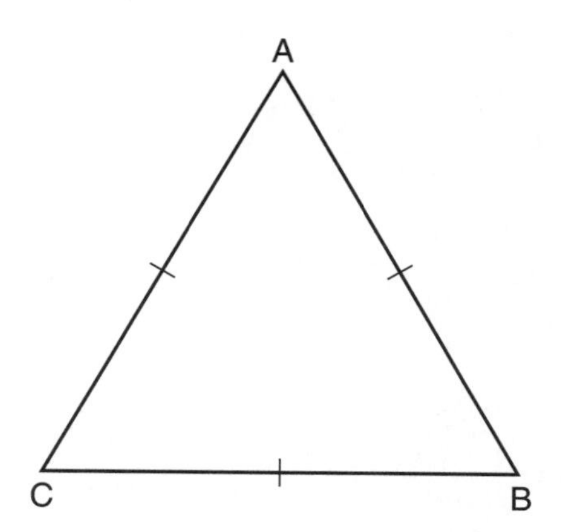

1. To what family of polygons does this shape belong?
2. What is the special name for this shape?
3. Which sides have the same length?
4. What do you know about angles A, B and C?
5. Side BA is opposite vertex C. Which side is opposite vertex A?
6. What is the size of angle B?
7. What name is given to angles B and C?
8. How many lines of reflection does the shape have?
9. What is the order of rotational symmetry of the shape?
10. How many diagonals does the shape have?
11. How many vertices does the shape have?
12. Suppose you cut off angle A with a horizontal cut.
    What is the name of the new polygon?
13. If you fold the corner of angle A down onto the point halfway along the side BC, what is the name of the new shape you have?
14. Which vertices would you have to cut off with a straight cut in order for the new polygon to be a hexagon?
15. Imagine a shape the same as ABC but bigger.
    Are the shapes similar or congruent?

P1.6HW Checking results

For all the following questions, remember to check your results.

1 **a** Find the sum of 255 and 217.

b Find the difference between 192 and 431.

c Calculate the product of 27 and 9.

2 Find x when $5x - 2 = 22$

3 $2m + 16 = 28$

Find m.

4 Four people measured the length of a playing field.

Which measurement is most appropriate?

97.4564 m	9546 cm	102.1 m	15 m

Explain why you chose your answer.

5 Copy and complete this magic square.

Every row, column and diagonal should add to the same total.

7		
	5	
		3

Can you fill it in with different numbers, to make another magic square?

Level 4

This table shows the distances between some towns.

Distances in miles

	Hull	Exeter	Bangor	Dover
Hull				
Exeter	305			
Bangor	199	289		
Dover	261	248	331	

a Which **two** towns are the **shortest distance** from each other? *1 mark*

b Mrs. Davis drove from Bangor to Exeter.

What is the distance between **Bangor** and **Exeter**? *1 mark*

Then Mrs. Davis drove from Exeter to Dover.

What is the distance between **Exeter** and **Dover**? *1 mark*

How far did Mrs. Davis drive altogether? *1 mark*

Level 5

a A football club is planning a trip.

The club hires **234** coaches. Each coach holds **52** pasengers.

How many passengers is that altogether?

Show your working. *2 marks*

b The club wants to put one first aid kit into each of the 234 coaches.

These first aid kits are sold in **boxes of 18**.

How many boxes does the club need? *1 mark*

A5.1HW Equations, formulae and functions

Remember:

You solve an equation by keeping it balanced.

$$3x + 2 = 17$$
$$3x = 17 - 2$$
$$3x = 15$$
$$x = 15 \div 3$$
$$x = 5$$

1 Solve these equations:

a $3m = 27$ **b** $4p - 10 = 22$ **c** $7t + 3 = 24$

2 **a** Write an equation for this problem:

Four packs of chewing gum and a 40p bar of chocolate cost 120p in total.

Hint: Use x for the cost of one pack of chewing gum.

b Solve your equation to find the cost of a pack of chewing gum.

3 Work out the values of the unknown in these diagrams:

a

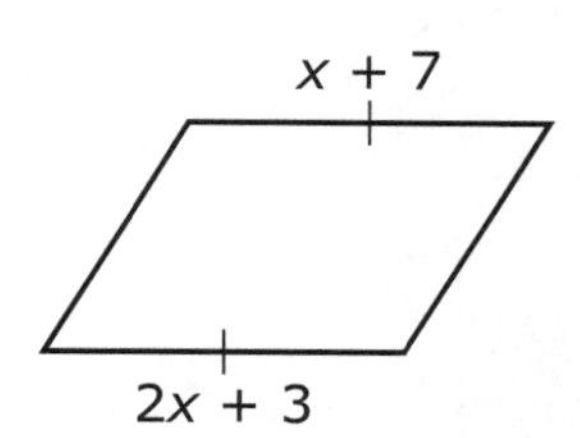

b

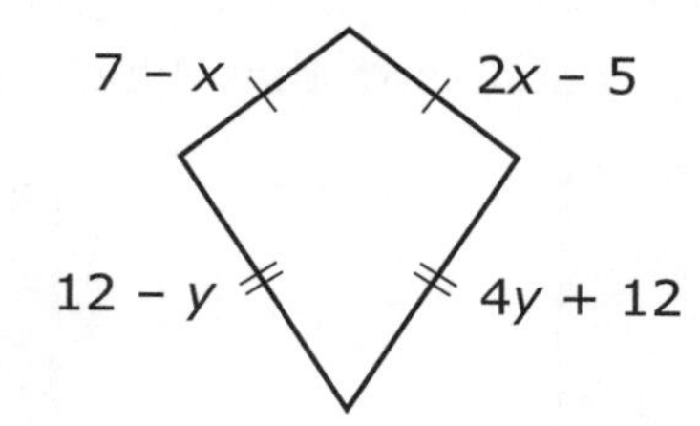

4 **a** Substitute the value $x = {}^{-}3$ into the expression $2x + 5$.

b Substitute the value $z = {}^{-}4$ into the expression $3z + 15$.

A5.2HW Using brackets

Remember:

- To expand brackets, you multiply each term inside the brackets by the term outside.

  $a(b + c) = ab + ac$

  $a(b - c) = ab - ac$

1 Expand the following brackets.

a $3(n + 4)$

b $5(x - 2)$

c $4(p + 6)$

d $^{-}2(e + 3)$

e $^{-}10(a + 3)$

2 Expand the brackets and simplify the expressions.

a $10 + 3(x + 2)$

b $2(p + 3) + 4(q + 4)$

c $3e + 3(e - 4) + 12$

d $4(x - 1) - 3x + 5$

e $4(x + 2y) - 2(2x + y)$

3 Find all the different expressions for the area of the shaded region in each of these shapes and show that they are all equivalent:

a

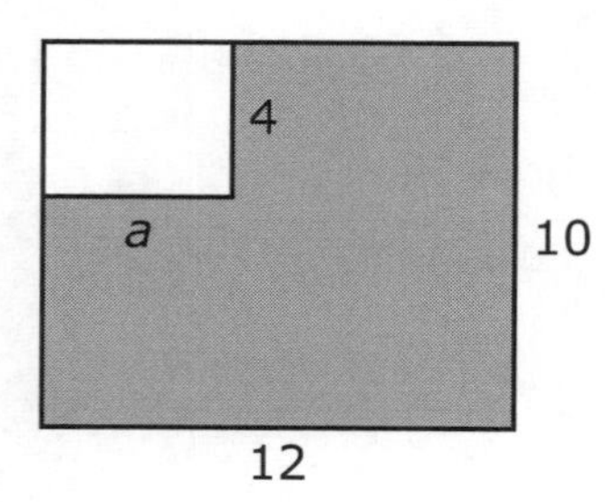

b

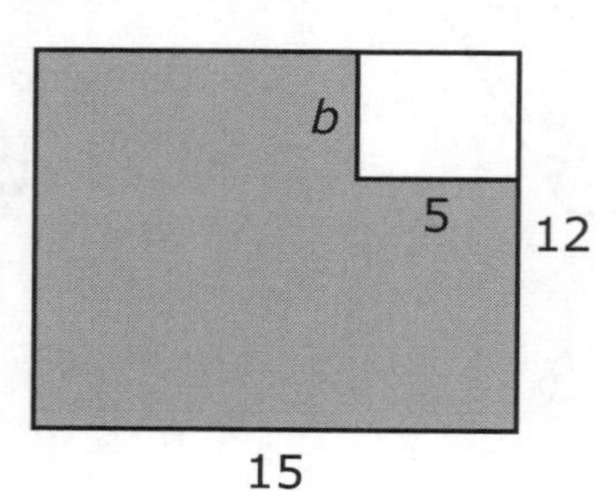

c

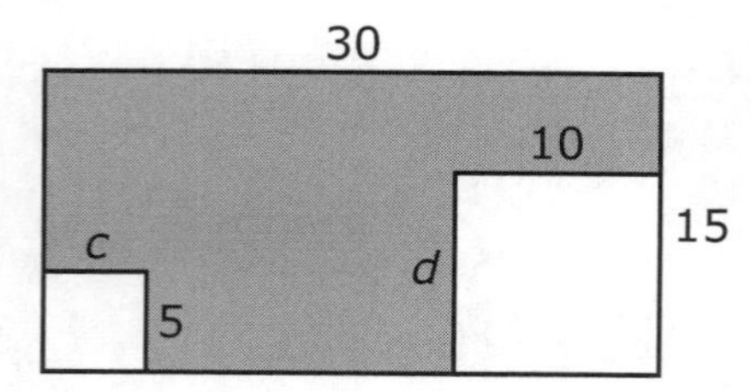

d

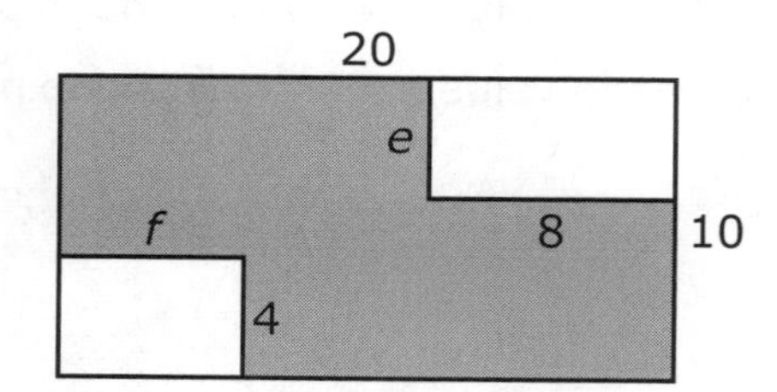

Cross number puzzle

Copy the grid.

Solve the puzzle using the solutions to the equations below.

Some of the clues have been completed for you.

a 1		*b* 2			*d*	*e*
	c 1	4		*f*		
g			*h*		*j*	
i	*m*	*n*			*k*	*q*
	p		*s*			
r					*v*	
t				*u*		

Across

$2a = 2$

$2(c - 4) = 20$

$d - 2 = 14$

$2h = 482$

$\frac{i}{2} = 312$

$2(k + 1) = 3k - 41$

$2p = 4480$

$t \div 11 = 11$

$4(u - 2) = 2(u + 10)$

Down

$2(b + 1) = 3b - 22$

$2(e + 6) = 140$

$5(f - 2) = 4(f + 1)$

$\frac{g}{2} = 8$

$7(j - 5) = 63$

$2m = 44$

$2(n + 3) = 90$

$4(q + 1) = 160$

$r^2 = 121$

$s \div 2 = 20.5$

$4(v + 1) = 6v - 44$

A5.4HW Equations involving fractions

Examples

Solve: **a** $\frac{x}{5} = 6$ **b** $\frac{12}{x} = 4$

a $\frac{x}{5} = 6$

$x = 5 \times 6$

$x = 30$

b $\frac{12}{x} = 4$

$12 = 4x$

$12 \div 4 = x$

$x = 3$

1 Solve each equation below and use this code to change your answers to letters.

A	B	C	D	E	F	G	H	I	J	K	L	M	N	O	P	Q	R	S	T	U	V	W	X	Y	Z
1	2	3	4	5	6	7	8	9	10	11	12	13	14	15	16	17	18	19	20	21	22	23	24	25	26

a $\frac{x}{2} = 6.5$

b $\frac{x+5}{6} = 5$

c $\frac{x}{2} = 3$

d $\frac{x+2}{3} = 1$

e $x - 2 = 20$

f $\frac{x+5}{2} = 10$

g $\frac{x}{7} = 3$

h $3(x + 2) = 60$

i $\frac{18}{x} = 2$

j $\frac{x-4}{2} = 8$

k $\frac{15-x}{2} = 5$

l $2(x + 3) = 10$

m $x - 1 = 0$

n $\frac{x+1}{3} = 5$

o $6(x - 2) = 12$

p $\frac{x}{3} = 3$

q $\frac{x+1}{4} = 5$

2 Make up your own questions to complete the sentence.

A5.5HW Using formulae

A company is designing new packaging for their new brand of sweets.

They need to work out which of the following cuboids gives the maximum volume.

1 **a** Using the formula for the volume:

$V = lwh$ where l = length, w = width, h = height

find the volume of each cuboid below:

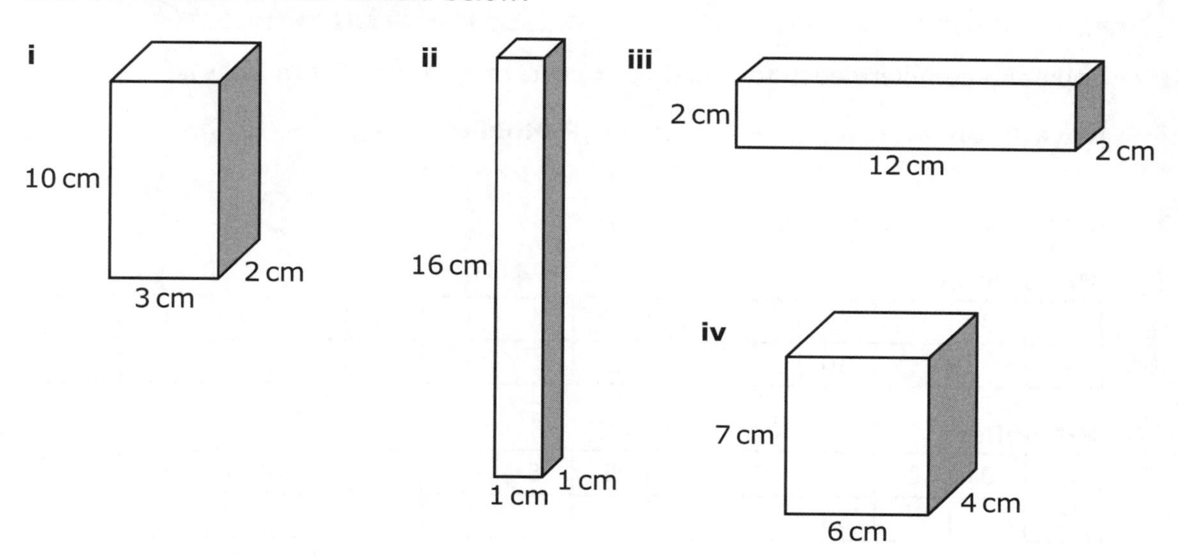

b Which box would you advise the company to use?

2 Work out the area of cardboard needed to make each of the boxes.

> **Hint**: Find the area of the net of each cuboid.

3 Have you changed your mind about which box you would advise the company to use? Explain your answer.

4 Write a formula for working out the surface area of a cuboid.

A5.6HW Solving equations using graphs

There are two special offers on the same phone that Emily likes but in two different shops.

NOVA PHONES

£30 to buy the phone *plus*
£10 per month for unlimited calls

P-Mobile

£15 to buy the phone *plus*
£12 per month for unlimited calls

1 Copy and complete each formula. (C = cost, m = number of months)

Nova Phones: C = ___ + ___ m **P-Mobile:** C = ___ + ___ m

2 Copy and complete the tables of values:

Nova Phones:

m	0	1	2	3	4	5	6	7	8	9	10	11	12
C			50										

P-Mobile:

m	0	1	2	3	4	5	6	7	8	9	10	11	12
C			39										

3 Draw a graph for each shop with both lines plotted on the same graph.

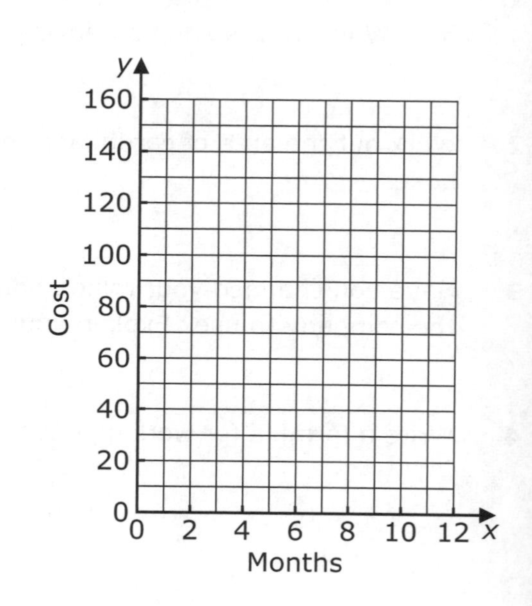

4 When do both mobile phone offers cost the same amount?

5 If you wanted to keep the mobile phone for only 6 months, would you buy it from Nova Phones or P-Mobile? Why?

6 If you wanted to keep the mobile phone for at least year, where would you buy it from? Why?

Level 4

This shape has 3 sides of length t and 2 sides of length s.

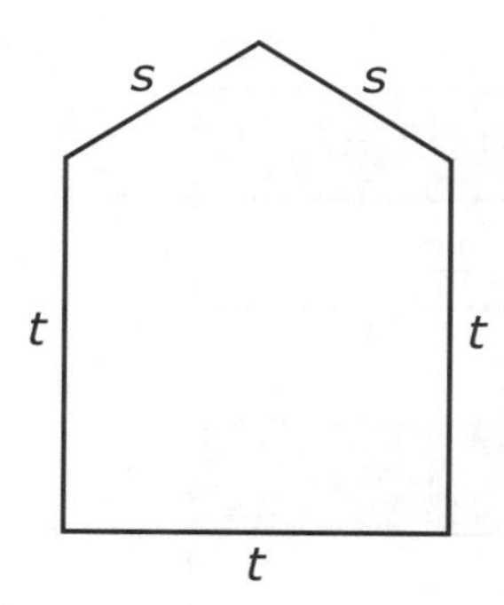

The perimeter of this shape is $3t + 2s$.

$$\mathbf{p = 3t + 2s}$$

Write an expression for the perimeters of each of these shapes.

Write each expression in its simplest form.

a

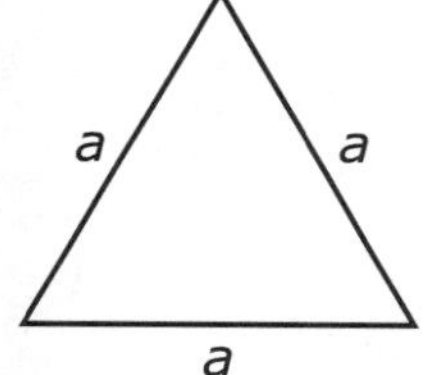

b

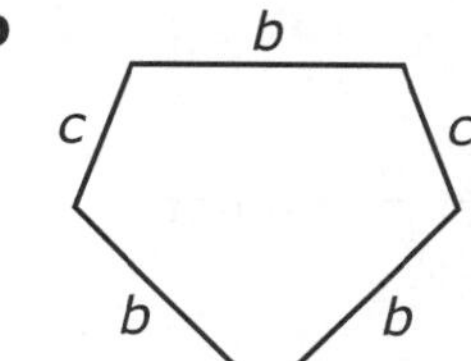

c

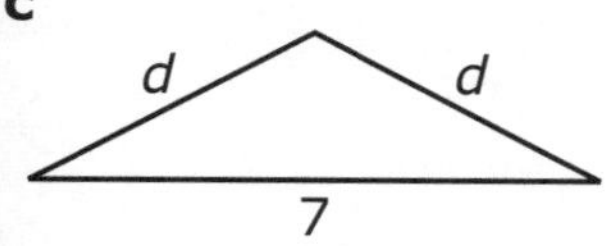

d

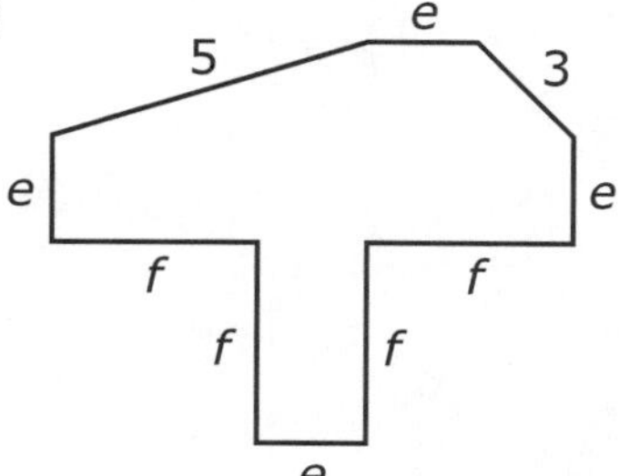

4 marks

Level 5

A cookery book shows how long, in minutes, it takes to cook a joint of meat.

Microwave oven
Time = (12 × weight in pounds) + 15

Electric oven
Time = (30 × weight in pounds) + 35

a How long will it take to cook a **3 pound** joint of meat in a **microwave oven**? *1 mark*

b How long will it take to cook a **7 pound** joint of meat in an **electric oven**? *1 mark*

c How much quicker is it to cook a **2 pound** joint of meat in a microwave oven than in an electric oven?

Show your working. *2 marks*

D3.1HW Collecting and organising data

Becky's class are starting work on a handling data project.
The theme is 'Transport'.
They have to choose their own questions to investigate.

Becky's teacher gave them this advice before starting:

Remember these things:

- Pick a question that is interesting and important.
- Try to turn your question into a specific hypothesis, which you can test.
- Plan the data collection. Decide which data you will need.
  Is it primary or secondary data?
  How will you collect and organise your data?
  How will you use the data?

Do **one** of the tasks described below.

Task 1

If you have started work on your own handling data project, write up a plan for your own project. (You may already have started on this in class.) You can use the notes above to help.

Task 2

If you have not started work on your own project, use the notes above to help you write a plan for a project on the theme of 'Transport'.

D3.2HW Calculating statistics

1 The frequency table below shows the results of a survey about the number of DVDs a group of 30 people rented from a video store in a month.

DVDs	0	1	2	3	4	5	6
Frequency	7	5	6	4	3	3	2

Work out the mean, median, mode and range of the number of DVDs rented.

> **Hint:** $\text{Mean} = \dfrac{\text{total number of DVDs}}{\text{total number of people}}$
>
> To work out the median (middle number), first write the number of DVDs in order of size:
> 0, 0, 0, 0, 0, 0, 0, 1, 1, 1, 1, 1, 2, 2, 2…

2 The heights of 30 students in a class are shown in the frequency table.

Height, h cm	Frequency
$150 \le h < 155$	2
$155 \le h < 160$	5
$160 \le h < 165$	10
$165 \le h < 170$	9
$170 \le h < 175$	4

a Which category would a height of exactly 160 cm be recorded in?

b Write down the modal class for the distribution.

c Copy and complete the frequency diagram for the data in the table.

Using an assumed mean

Remember:

- To find a mean using an assumed mean:
  - Assume a mean.
  - Find the difference of the assumed mean from each data value.
  - Find the mean of the differences.
  - Add the assumed mean back on.
- You can use the formula:

$$\text{Mean} = \text{Assumed mean} + \frac{\text{Total number of differences}}{\text{Number of items}}$$

1 The table shows the age at which 12 people passed their driving test.

Age (years)	18	17	25	30	19	17	18	22	21	18	20	21
Difference from 17	1	0										

a Copy and complete the table to show the differences from 17 years.

b Work out the mean age of the 12 people.

2 Adam recorded the times taken (in seconds) for 12 people to complete a task.

In the second row of the table he started to work out the difference from 60 seconds for each time.

Time (s)	70	47	75	61	63	50	62	59	47	60	69	64
Difference from 60 s	$^{+}10$	$^{-}13$	$^{+}15$							0		

a Copy and complete the table. Work out all the differences from 60 seconds.

b Work out the sum of the positive differences:

$10 + 15 + \ldots$

c Work out the sum of the negative differences:

$^{-}13 + \ldots$

d Add your answers to **b** and **c** to find the total of **all** the differences.

e Use this formula to calculate the mean time:

$$\text{Mean time} = 60 + \frac{\text{Total of all the differences}}{\text{Number of items}}$$

D3.4HW Graphs for data

1 Jamila grew a sunflower.

She recorded its height every two weeks, for 20 weeks.

Week	2	4	6	8	10	12	14	16	18	20
Height (cm)	4	10	16	30	55	79	102	140	170	190

a Draw a set of axes with 'Weeks' on the horizontal axis and 'Height' in the vertical axes. (The horizontal axis goes from 0 to 20 weeks. The vertical axis goes from 0 to 190 cm.)

b Plot the data from the table on your axes.

c Join the points with a smooth curve to make a line graph.

2 **a** In a survey, students recorded the number of hours in a week they spent reading and playing computer games.

Reading (hours)	3	1	2	10	1	8	5	3	4	3
Computer games (hours)	1	5	6	2	3	3	2	10	5	7

Plot a scatter graph to show this data.

Copy and complete this sentence:

The more time students spend on __________ the less time they spend on ________________.

b In the same survey, students recorded the number of chocolate bars they ate.

Reading (hours)	3	1	2	10	1	8	5	3	4	3
Chocolate bars	1	3	2	4	2	3	4	6	3	2

Plot a scatter graph to show this data.

From your graph, do you think the amount of time spent reading is related to the number of bars of chocolate eaten?

D3.5HW Stem-and-leaf diagrams

1 As part of her survey on Transport, Becky recorded the number of people on one side of every bus that passed in 30 minutes on a Saturday morning.

11, 12, 14, 12, 13, 7, 4, 12, 5, 13, 12, 4, 13 12, 13, 12, 11, 4, 6, 11, 12, 15, 14, 2, 1

a Show this data on a stem-and-leaf diagram.
Remember to order the data first.
Don't forget to show the key.

b Use your stem-and-leaf diagram to find

i the range of the number of people on one side of a bus.

ii the modal number of people on one side of a bus.

2 Becky recorded the number of people on one side of every bus that passed on a Monday morning.

11, 12, 12, 11, 11, 3, 4, 12, 11, 4, 5, 3, 12, 11, 9, 3, 2, 11, 10, 5

a Draw a stem-and-leaf diagram for this data.

b Use your diagram to find the modal number of people on one side of a bus.

3 Compare your answers to questions **1** and **2**.
What is the modal number of people on one side of a bus on

a Saturday morning

b Monday morning?

Suggest a reason for this.

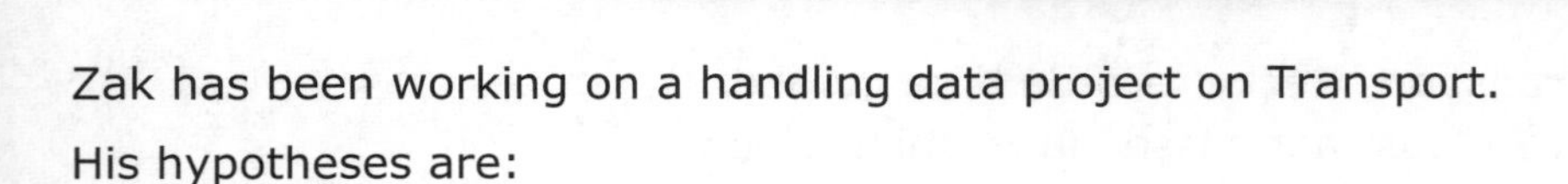

Communicating results

Zak has been working on a handling data project on Transport.

His hypotheses are:

- In Year 9, more boys than girls cycle to school.
- In Year 9, more girls than boys are taken to school by car.

Zak surveyed the Year 9 students.

The table shows his results.

	Method of transport to school				
	Walk	Cycle	Car	Bus	Other
Girls	45	30	20	75	10
Boys	35	42	25	80	8

Write a conclusion for Zak's project. It should be about a paragraph in length.

- Include a graph or chart to show the results of the survey.

  Remember to think carefully about what sort of graph or chart to use.

  Examples could be: pie chart, bar graph, bar-line graph, etc.

- Explain whether or not the results confirm the hypotheses.

  Remember to consider each one separately.

Level 4

There are **24 pupils** in Jim's class.

He did a survey of how the pupils in his class travelled to school.

He started to draw a pie chart to show his results.
Copy the pie chart.

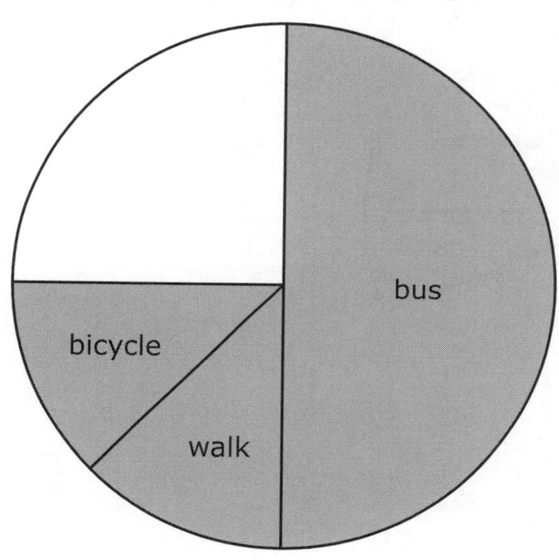

a **4** pupils travelled to school by **train**.

Show this on Jim's pie chart as accurately as you can.

Label this part **train**.

Label the remaining part **car**. *1 mark*

continued

Level 4 *Continued from page 111*

b There are **36 pupils** in Sara's class.

She did the same survey and drew a pie chart to show her results.

15 pupils travelled by **bus** and 6 pupils **walked**.

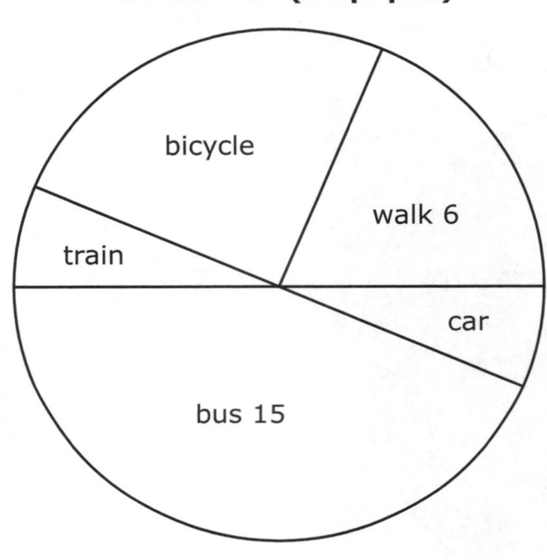

How many pupils travelled to school:

i by train

ii by car

iii by bicycle? *2 marks*

c Jim says:

> **15** pupils in Sara's class travelled by **bus**.
> Only **12** pupils in my class travelled by **bus**.
> Sara's pie chart shows **fewer** people
> travelling by bus that mine does.
> So Sara's chart must be wrong.

Explain why Jim is wrong. *1 mark*

Level 5

Some pupils wanted to find out if people liked a new biscuit.
They decided to do a survey and wrote a questionnaire.

a One question was:

How old are you (in years)?

☐ 20 or younger ☐ 20 to 30 ☐ 30 to 40 ☐ 40 to 50 ☐ 50 or older

Mary said:

The labels for the middle three boxes need changing.

Explain why Mary was **right**. *1 mark*

b A different question was:

How much do you usually spend on biscuits each week?

☐ a lot ☐ a little ☐ nothing ☐ don't know

Mary said:

Some of these labels need changing too.

Write new labels for any boxes that need changing.

You may change as many labels as you want to. *2 marks*

continued

Level 5 *Continued from page 113*

The pupils decide to give their questionnaire to 50 people.

Jon said:

c Give **one disadvantage** of Jon's suggestion. *1 mark*

d Give **one advantage** of Jon's suggestion. *1 mark*

S4.1HW Tessellations

1 Copy these shapes onto squared paper.
Show how each shape tessellates, using translations and rotations.

a

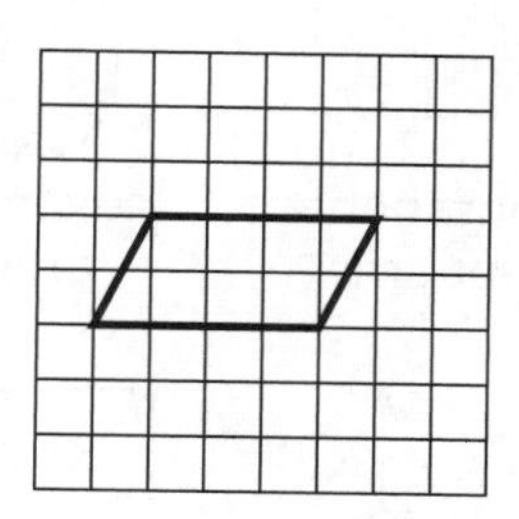

b

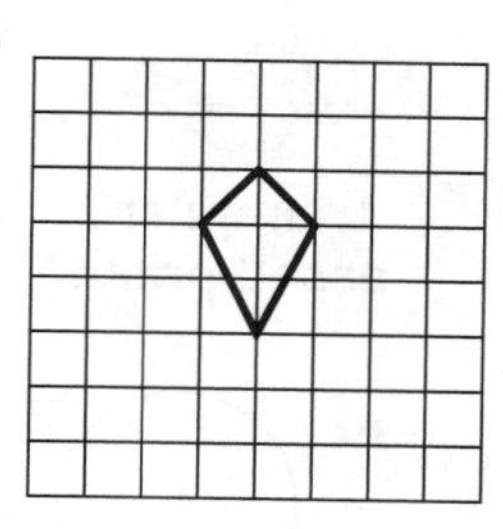

c

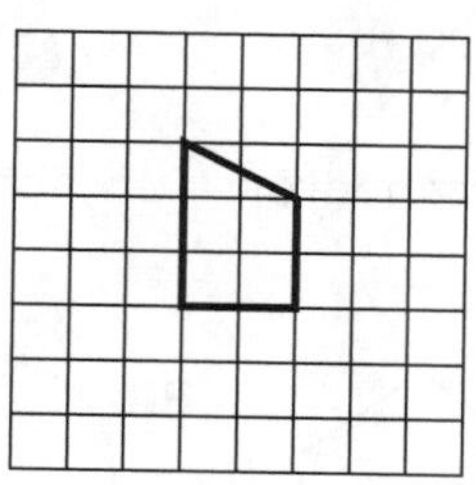

d

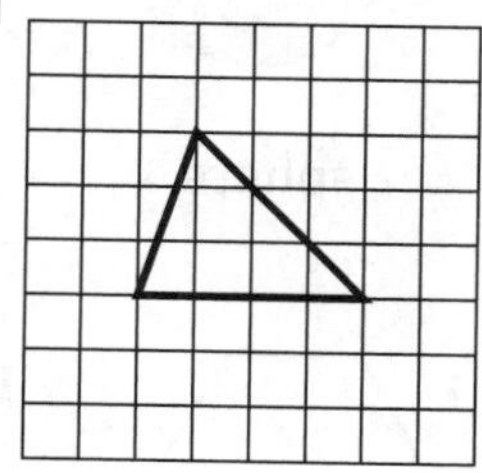

2 **a** Construct accurately and cut out three triangles with these measurements.
Mark the angles on both sides of the paper.

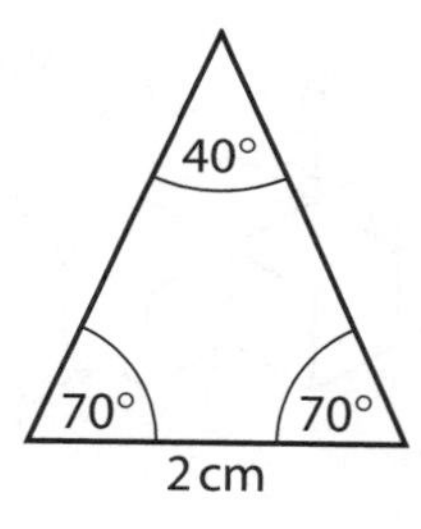

b Rearrange two of the triangles to form a rhombus, a parallelogram and a kite.

c In each case, draw the outline and mark the angles.

d Arrange all three triangles to form an isosceles trapezium.

e Draw the outline and mark the angles.

f Give reasons why it is possible to make each quadrilateral.

S4.2HW Plans and elevations

1 Match each of these solids with its plan view.

a 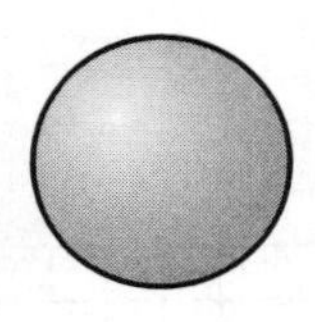
sphere

b octahedron

c cone

d 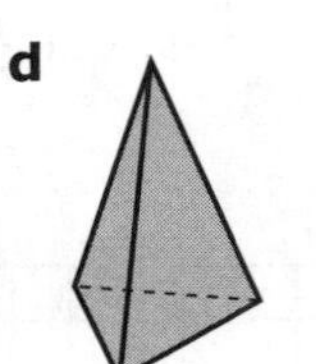
triangular-based prism

e 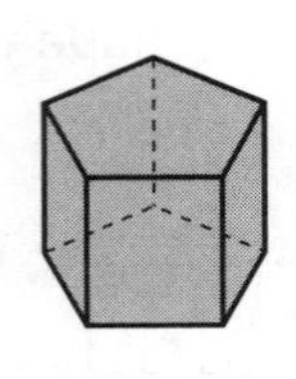
pentagonal prism

f 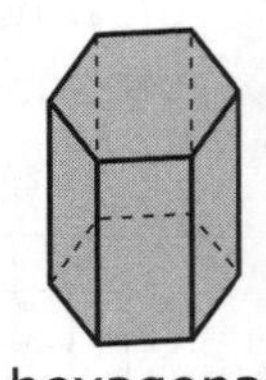
hexagonal prism

i

ii 

iii 

iv 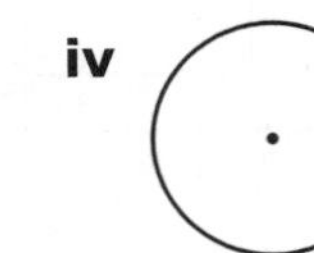

v 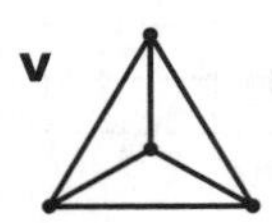

vi 

2 On squared paper, draw the plan, the side elevation and the front elevation of each shape.

A 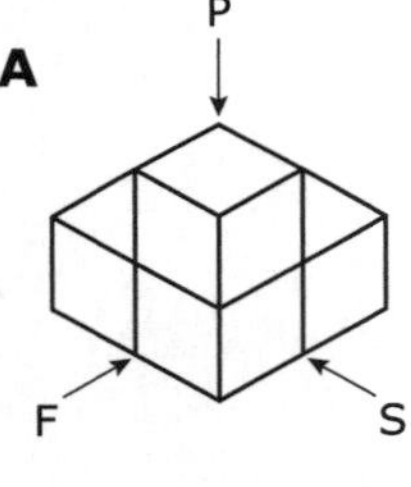

B 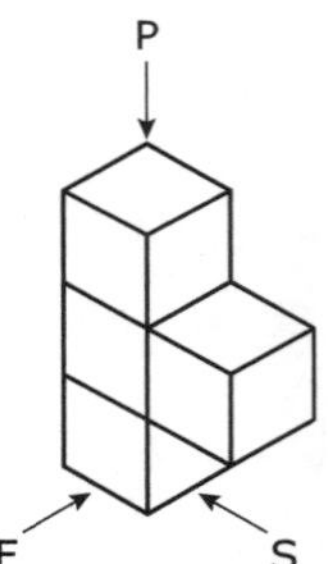

C 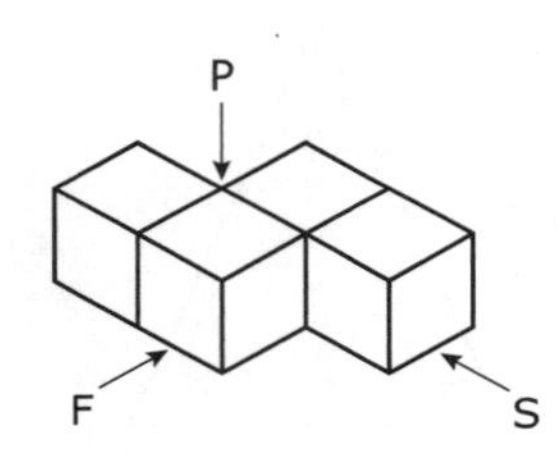

D 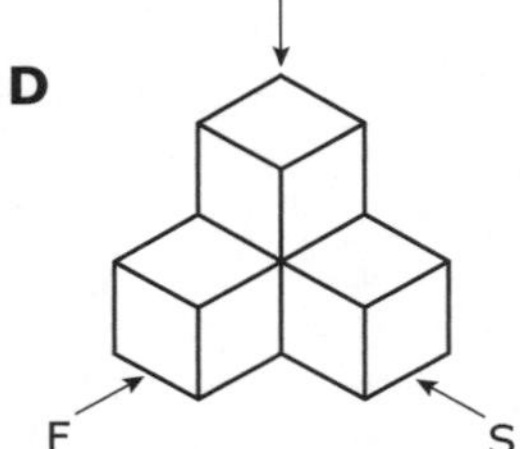

3 **a** Draw each 3-D shape on isometric paper, given the three views below.

b How many cubes are needed to make each shape?

A

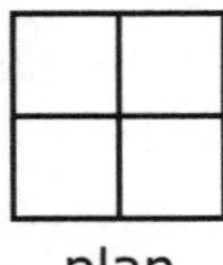
plan

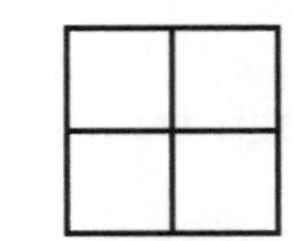
side elevation

front elevation

B

plan

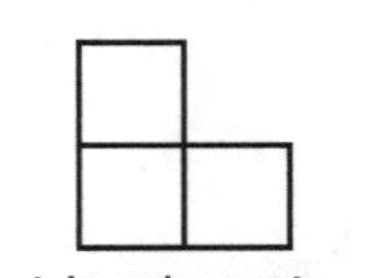
side elevation

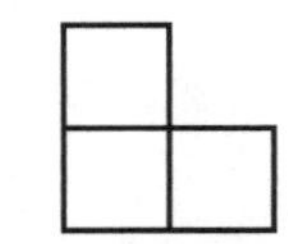
front elevation

S4.3HW Volume of a cuboid

Remember:

- Volume of cuboid = length × width × height

1 **a** Calculate the volume of these cubes, then continue up to a 10 cm by 10 cm by 10 cm cube.

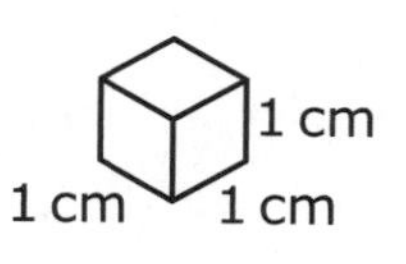

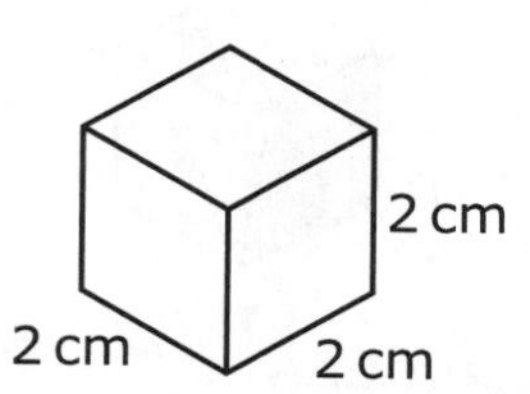

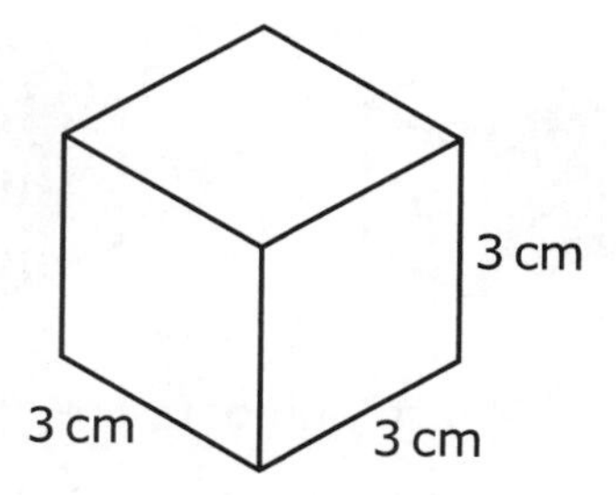

b List your answers in a sequence.
What is the name of these numbers?

2 32 cubes are put together to make a cuboid.

a Give the dimensions of the five possible cuboids that could be built.

b Which of the cuboids is the nearest to a cube?

3 Calculate the volume of these shapes.

a

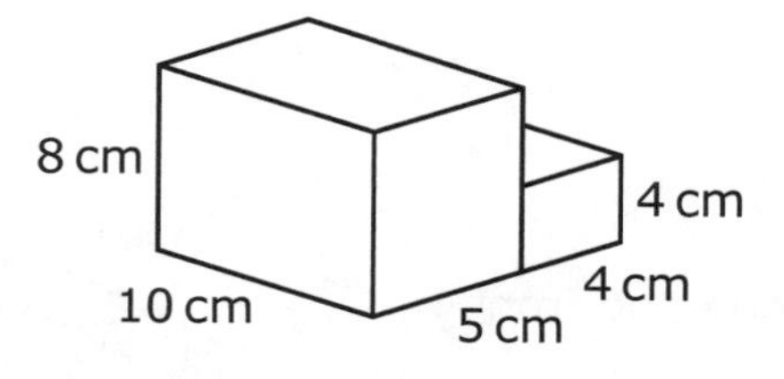

b

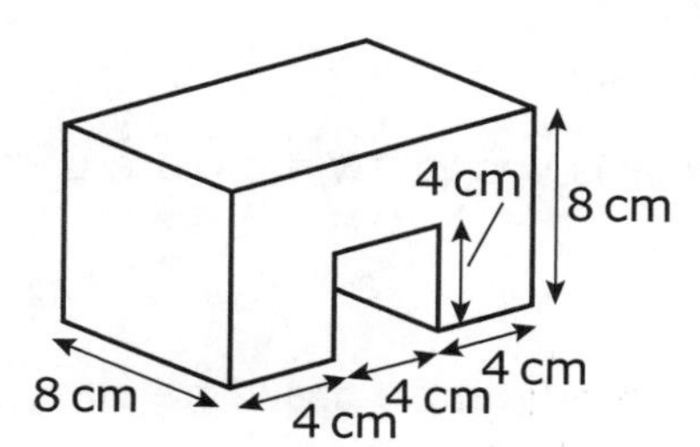

4 How many of these cartons of orange juice will fit into the box?

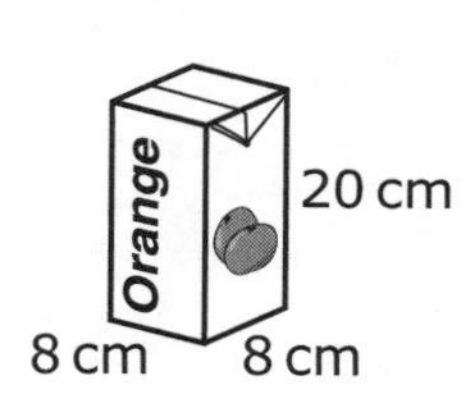

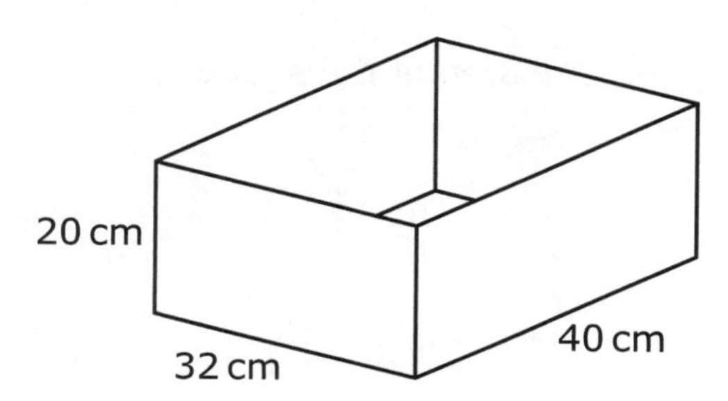

S4.4HW Surface area of a cuboid

Remember:

- The surface area of a cuboid is the area of its net.

1 A cube has six different numbered faces.

Here are three different views of the same cube.

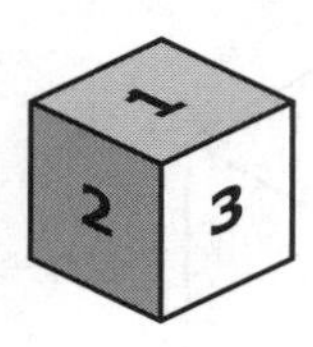

a Which numbers are opposite each other?

b Write the numbers in the correct place on a copy of this net of a cube.

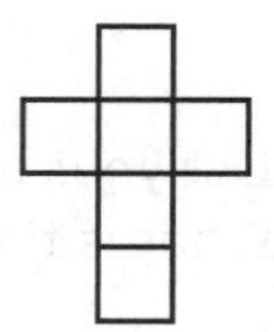

2 A one metre cube is put on the floor in front of you.

Can you position yourself to view

a one face

b two faces

c three faces

d four faces?

If the answer is yes, draw a picture of your view.

3 **a** On squared paper, draw the plan, side elevation and front elevation of this shape.

b Calculate the area of the plan, side elevation and front elevation.

c Calculate the surface area of the shape.

S4.5HW Nets and scale drawings

1 **a** Copy these nets on to squared paper and cut them out.

Name the solid that can be formed from each net.

i

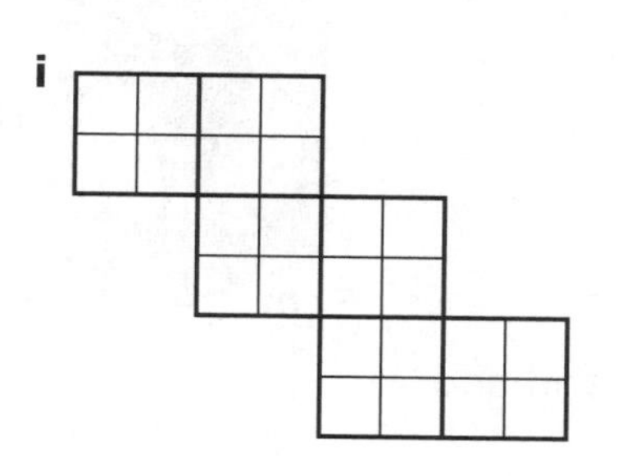

ii

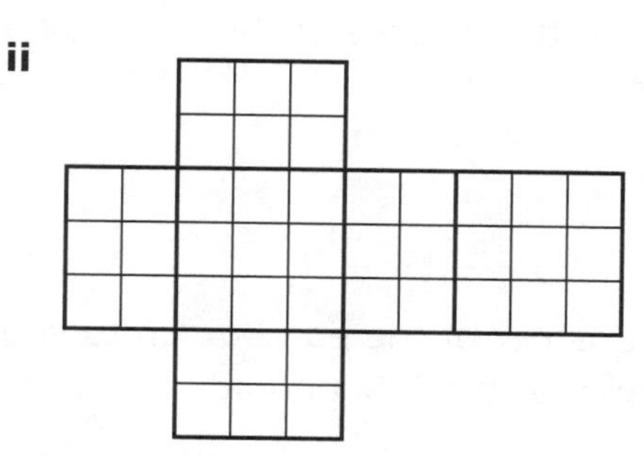

b Now redraw the nets on squared paper, but add flaps in the appropriate places so that a box with a lid can be made from each net.

2 Together these nets will make a scale model of a garden shed.

The scale of the model is 1:100.

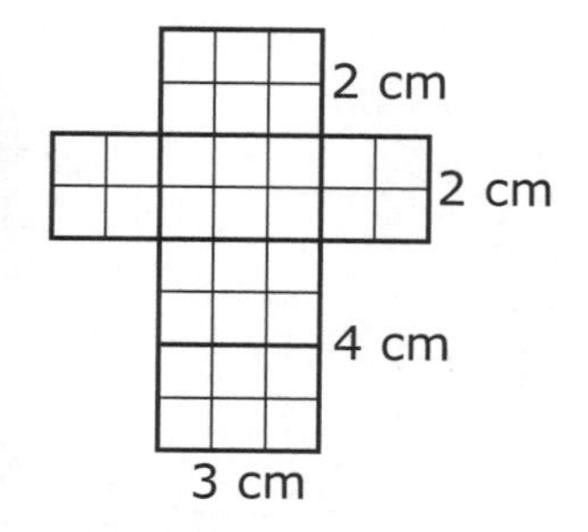

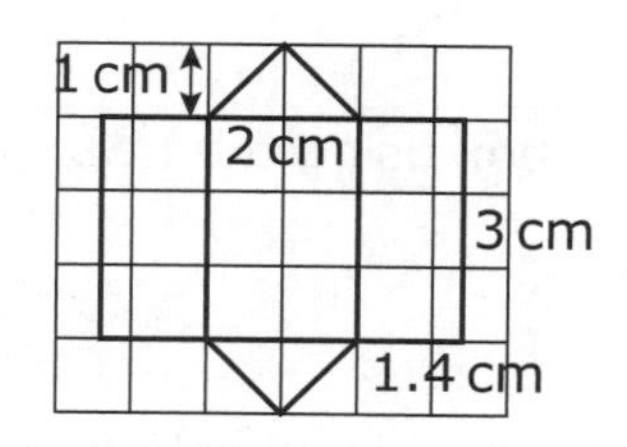

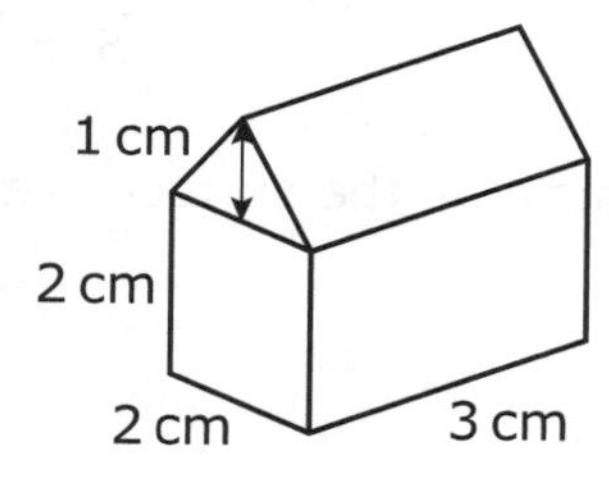

a What is the length, width and overall height of the real shed?

b Draw the two nets, and add flaps as needed.
Cut out the nets and make the model shed.

c Calculate the surface area in square metres of the sloping roofs for the real shed.

d Calculate the surface area in square metres of the other visible faces (do not include the base of the cuboid).

e Calculate the volume in cubic metres for the real-sized shed.

S4.6HW Designing a box

1 A chocolate egg is to be packaged in a cuboid.

a Explain why a cuboid is a good shape for packaging.

b Calculate the volume of this 5 cm by 5 cm by 10 cm cuboid.

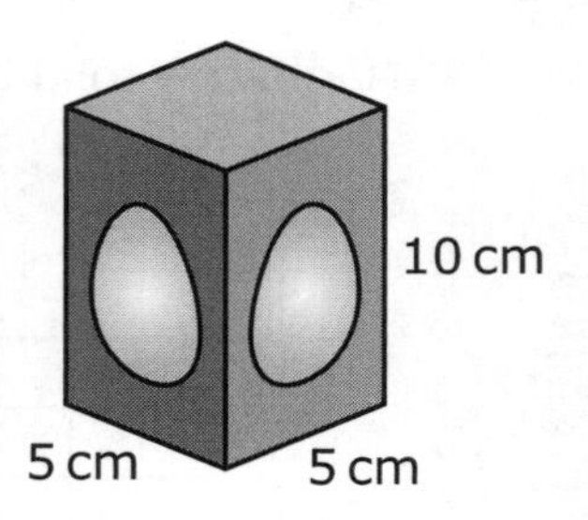

2 **a** Copy this net of the egg box on to squared paper and add the flaps for the box and lid.

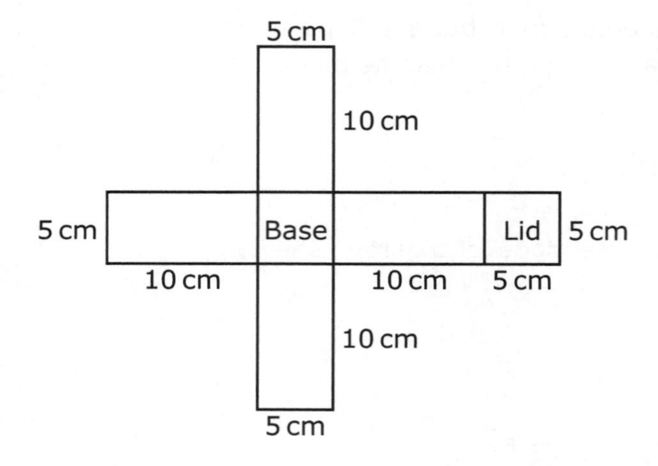

b Calculate the surface area of the box using the net.

3 **a** Sketch three views of the box: plan, side elevation and front elevation. Label the measurements on your sketches.

b Make an accurate full-sized drawing of the front elevation.

On your drawing, make a design for the front of the chocolate egg box.

Level 4

This is a right-angled triangular tile:

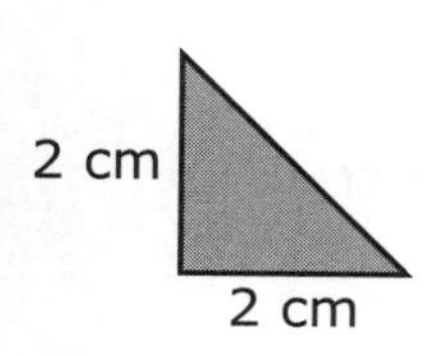

You can fit **8** of the tiles into a 4 cm by 4 cm square like this:

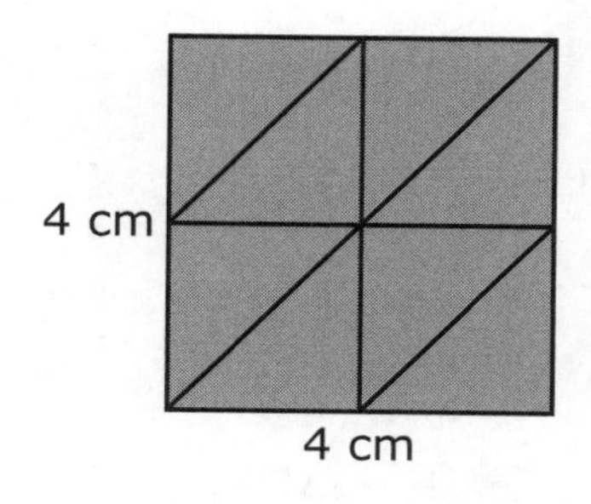

Write **how many** of the tiles you can fit into each of these shapes.

a

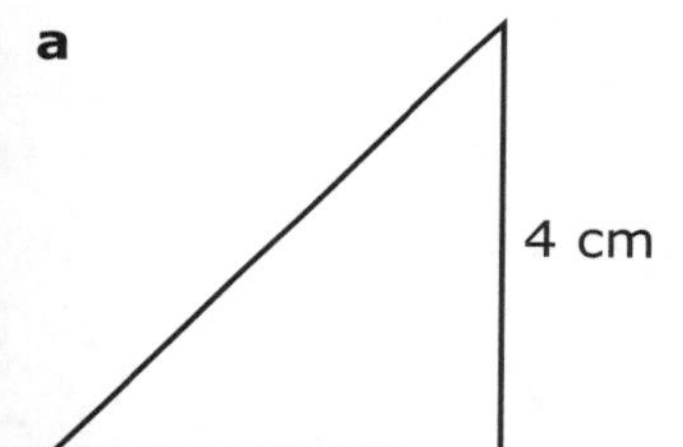

b

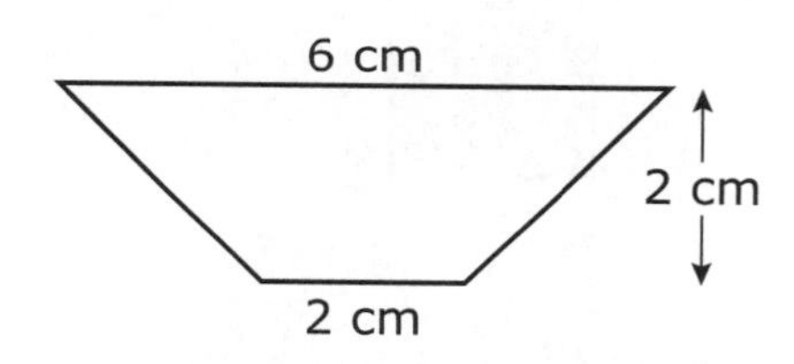

2 marks

c

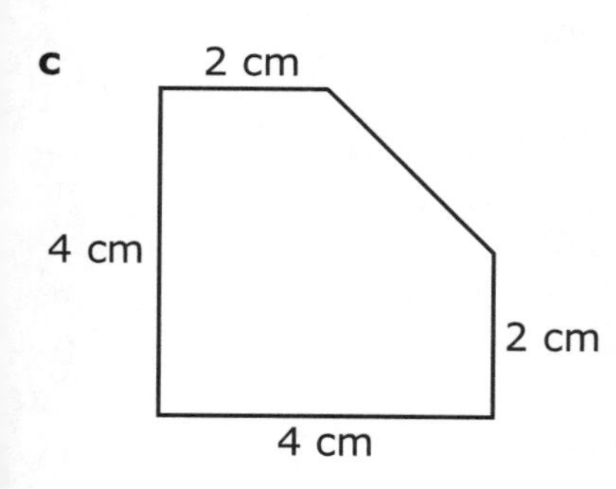

d

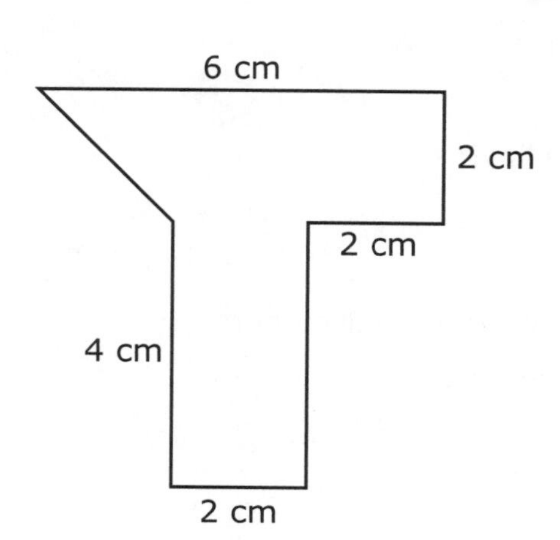

2 marks

Level 5

The sketch shows the net of a triangular prism.

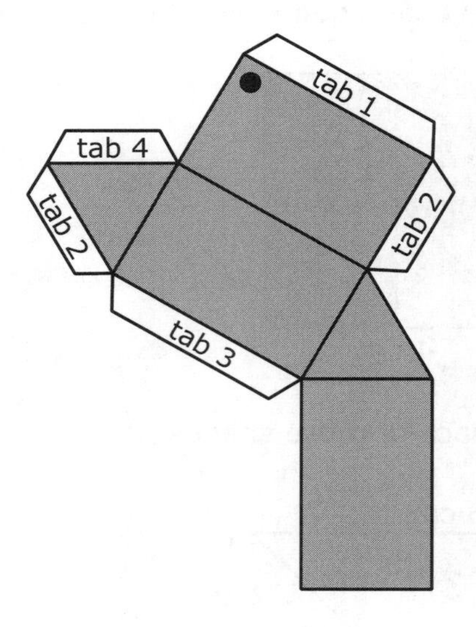

Copy the diagram.

The net is folded up and glued to make the prism.

a Which edge is **tab 1** glued to?

On your diagram, label this edge A. *1 mark*

b Which edge is **tab 2** glued to?

Label this edge B. *1 mark*

c The corner marked • meets two other corners.

Label these two other corners • *1 mark*

D4.1HW Simple tree diagrams

Remember:

- If the probability that an event occurs is p, then the probability that the event does not occur is $1 - p$.

Draw a tree diagram to represent each of these situations:

1 A fair coin is tossed, and the outcome is either Heads or Tails.

2 A counter is chosen at random from a bag containing equal numbers of red, blue and green counters.

3 A square spinner has sections marked A, B, C and D. The probabilities of getting each outcome are equal.

4 A triangular spinner has sections marked A, B and C. The probability of getting A is 0.3, and the probability of getting B is 0.5.

5 A bag contains 24 red sweets, 36 blue sweets and 40 black sweets. A sweet is chosen from the bag at random.

6 A counter is chosen at random from a bag. Half the counters in the bag are orange, 24% are blue and the rest are green.

1 Ice cream comes in three flavours: vanilla, banana and strawberry.

There is a choice of sauce: toffee or chocolate.

a Copy and complete the tree diagram to show all the possible combinations.

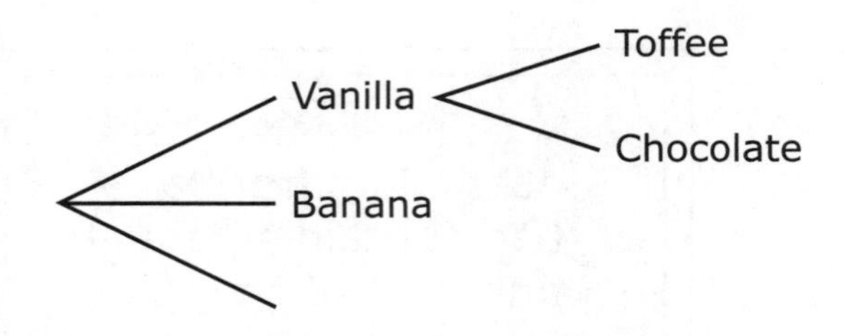

b Copy and complete this two-way table to show all the possible combinations.

	Vanilla	Banana	Strawberry
Toffee			(T, S)

2 This two-way table shows all the possible outcomes when two coins are tossed.

		Coin 1	
		Head	Tail
Coin 2	Head	H, H	H, T
	Tail	T, H	T, T

Show this information as a tree diagram.

Equally likely outcomes

1 This spinner has sections numbered 1 to 5.

The probabilities of the outcomes are equal.

a Draw a tree diagram to show the possible outcomes.

b Mark the probabilities on your tree diagram.

2 Another spinner has three coloured sections: Red (R), Green (G) and Blue (B).

Each colour is equally likely.

Work out the probability of each outcome.

3 The spinner from question **1** is spun, and then the spinner from question **2**.

a Copy and complete the tree diagram to show all the possible outcomes.

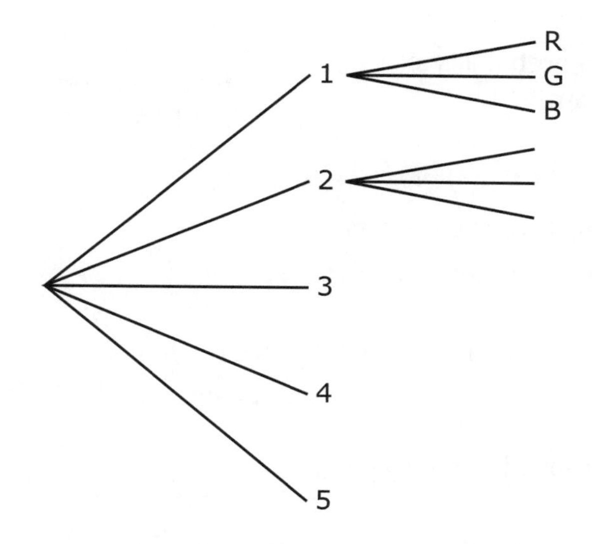

b How many possible outcomes are there?

c How many possible outcomes are there for an even number and Red together?

d What is the probability of an even number and Red together?

Hint: Use your answers to **b** and **c**.

D4.4HW Experimental probabilities

The diagram shows a river which branches on its way to the sea.
There are six places, A to F, where the river reaches the sea.

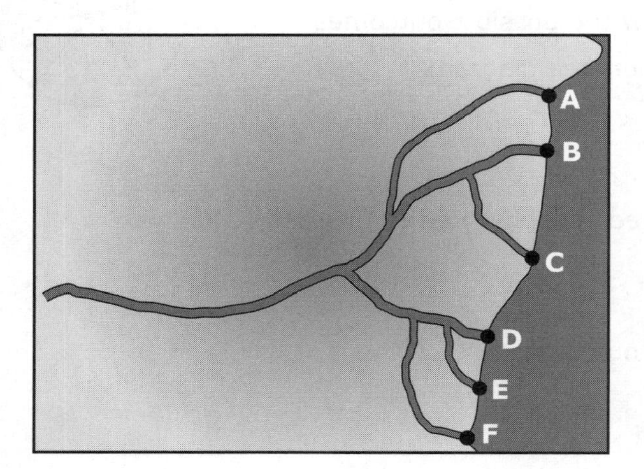

1 Make up a river diagram of your own. You may find it useful to draw a diagram with straight lines.

2 Write down your predictions for the probability of getting to each town on your diagram.

> **Hint:** The probabilities should add up to 1.

3
- ◆ Move a counter down the river, towards the sea.
- ◆ Every time you reach a fork in the river, toss a coin to decide which branch to follow.
- ◆ Record the town that you get to each time in a tally chart.

4 Use this formula to estimate the probability of reaching each town, from your results:

$$\text{Probability of reaching a town} = \frac{\text{Number of trips that reach the town}}{\text{Total number of trips down the river}}$$

D4.5HW More experimental probabilities

The diagram shows the possible routes from a starting point to three different finishing points, A, B and C.

At each 'fork' there is a probability of 0.3 of going straight on, and a probability of 0.7 of turning off to the right.

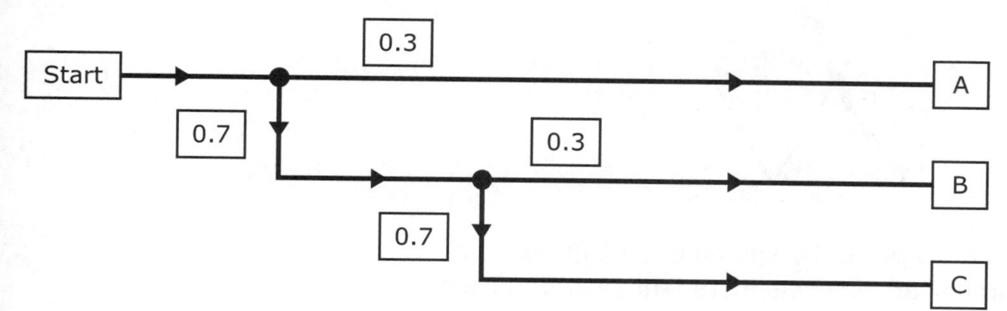

1 Carry out an experiment to estimate the probability of getting to each finishing point.

Place a counter at the start.

- Use some of the instructions in the table to decide whether to go straight on or turn right each time.

  These directions were generated by computer. Start anywhere you like in the table, and read it in any direction. R means 'turn right' and S means 'straight on'.

R	R	R	R	R	R	S	R	R	R
S	R	S	R	S	S	S	R	S	R
R	S	R	R	R	R	R	R	S	R
S	S	R	R	S	R	R	R	R	S
R	R	R	R	R	S	R	R	S	S
S	R	R	R	R	S	R	R	S	S
R	R	S	R	R	R	R	R	S	R
R	R	R	R	R	R	S	R	R	R
R	R	S	R	S	R	S	R	R	R
R	R	R	S	S	R	R	R	S	S

- Record where you finish in a tally chart.

2 When you have carried out a large number of trials, estimate the experimental probability of getting to each destination.

> **Hint:** Use the formula:
>
> $$\text{Probability of finishing at the point} = \frac{\text{Number of trips that reach the point}}{\text{Total number of trips}}$$

D4.6HW Comparing experimental and theoretical probabilities

This spinner was tested in an experiment.
The spinner is not fair – 'win' and 'lose' are not equally likely.

30 students each tested the spinner by spinning it 10 times.
The table shows the number of 'wins' each student recorded from their 10 spins.

6	2	4	6	5	4	3	5	4	2
4	2	1	4	3	2	4	5	5	1
4	4	4	5	5	6	4	3	3	4

1 **a** What was the largest estimated probability of getting a 'win' that a student could have calculated from their results?

Hint: What is the largest number of wins recorded?

b What was the smallest?

Explain your answers. How reliable do you think the estimated probabilities would be?

2 Make a more reliable estimate of the experimental probability, by combining the results for the whole class.

Estimate the probability of a win by calculating:

$$\frac{\text{Total number of wins}}{\text{Total number of trials}}$$

D4 Probability experiments

SAT QUESTIONS

Level 4

a Joe has these cards:

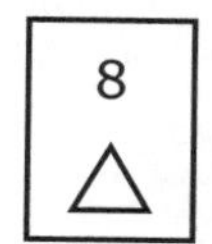
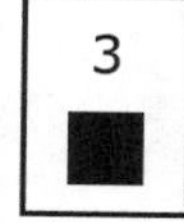
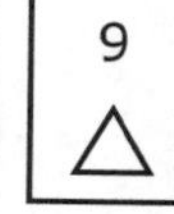

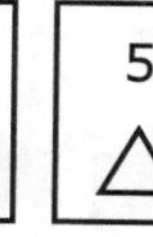
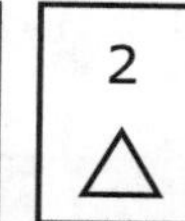
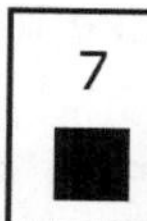

Sara takes a card without looking.

Joe says:

i Explain why Joe is **wrong**. *1 mark*

Here are some words and phrases:

impossible	not likely	certain	likely

Write a word or phrase that will complete these sentences.

ii It is _______ that the number on Sara's card will be **smaller than 10**. *1 mark*

iii It is _______ that the number on Sara's card will be an **odd number**. *1 mark*

continued

Level 4 *Continued from page 129*

b Joe mixes his cards and puts them face down on the table.

Then he turns the first card over, like this:

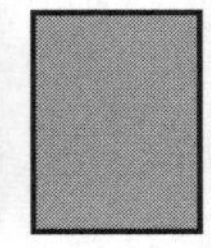

Joe is going to turn the next card over.

i Copy and complete this sentence:

On the next card, _____ is **less likely** than _____ *1 mark*

ii The number on the next card could be higher than 5 or lower than 5.

Which is **more likely**?

Select one of these statements and write it down.

higher than 5 *lower than 5* *cannot tell*

Explain your answer. *1 mark*

Level 5

Karen and Huw each have three cards, numbered 2, 3 and 4.

They each take any **one** of their own cards.

They then **add** together the numbers on the two cards.

The table shows all the possible answers.

	+	2	3	4
Huw	2	4	5	6
	3	5	6	7
	4	6	7	8

(Karen: columns; Huw: rows)

a What is the **probability** that their answer is an **even** number? *1 mark*

b What is the **probability** that their answer is a number **greater than 6**? *1 mark*

c Both Karen and Huw still have their three cards, numbered 2, 3 and 4.

They each take any one of their own cards.
They then **multiply** together the numbers on the two cards.

Draw a table to show all possible answers. *1 mark*

d Use your table in part **c** to write down the missing number in each of these statements.

The probability that their answer is a number that is less than ______ is $\frac{8}{9}$. *1 mark*

The probability that their answer is a number that is less than ______ is **zero**. *1 mark*

B1.1HW Starting an investigation

In your lesson, you started to investigate the cost of different rectangles made with rods and connectors.

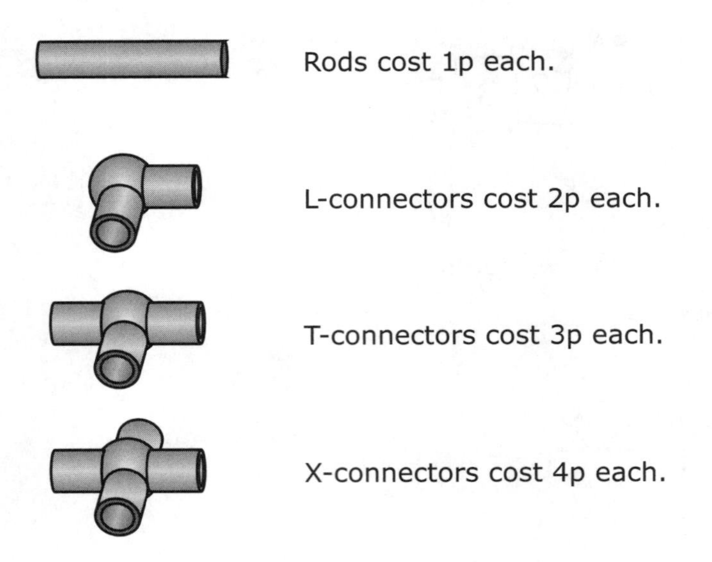

The challenge in the lesson was to find the largest rectangle that you can build with this kit for £1 or less.

- If you need to, you could do more work on finding the largest rectangle.

  Make sure that you have tried all of the possibilities, and record your work carefully.

- If you have already found the largest rectangle you could make for £1, you could take the investigation further by finding the largest rectangle that you can make for £2.

  Record your results carefully, so that you can continue later without repeating anything or leaving anything out.

B1.2HW Planning and recording

In the lesson, you looked at the importance of working systematically and putting the results of your investigation into tables.

What you need to do now depends on how well you have organised your results so far.

Do one of these tasks:

1 If your results are already properly organised in tables.

a Look for patterns in your results so far.

b Use the patterns to check the values in your table.
If you check and correct a result because of a pattern you spot in a table, you should mention this in your final project. Do not just throw away your 'mistakes' – showing how you put them right can be an excellent way to demonstrate some good mathematical thinking!

c Make some predictions, using the patterns in your table.

For example, predict the cost of 4 rectangles with

i height 4, width 3 **ii** height 2, width 13.

2 If your results are not properly organised.

Make some tables for your results so far.

		Width of rectangle											
		1	2	3	4	5	6	7	8	9	10	11	12
Height	1												
	2												
	3												

- Add as many rows and columns as you need.
- Fill in as many results as you can.
- You should find that there are number patterns in the results, and you can use these patterns to help you fill in the table.

B1.3HW Finding patterns and rules

You should now be at the stage where you can find and describe patterns in your results.

In the lesson:

- You looked at how a pattern of rectangles 'grew'.
  You found out which pieces you needed to add to turn each rectangle into the next biggest one.
- You saw how this pattern in the series of rectangles resulted in a pattern in the numbers in the results table.
  'Adding the same pieces each time' is the same as 'adding the same amount of money each time'.
- You used algebra to find a rule connecting the width of a rectangle W to its cost C.

You should continue with your work from the lesson, finding the number patterns in your results.

The flow chart shows the steps to follow:

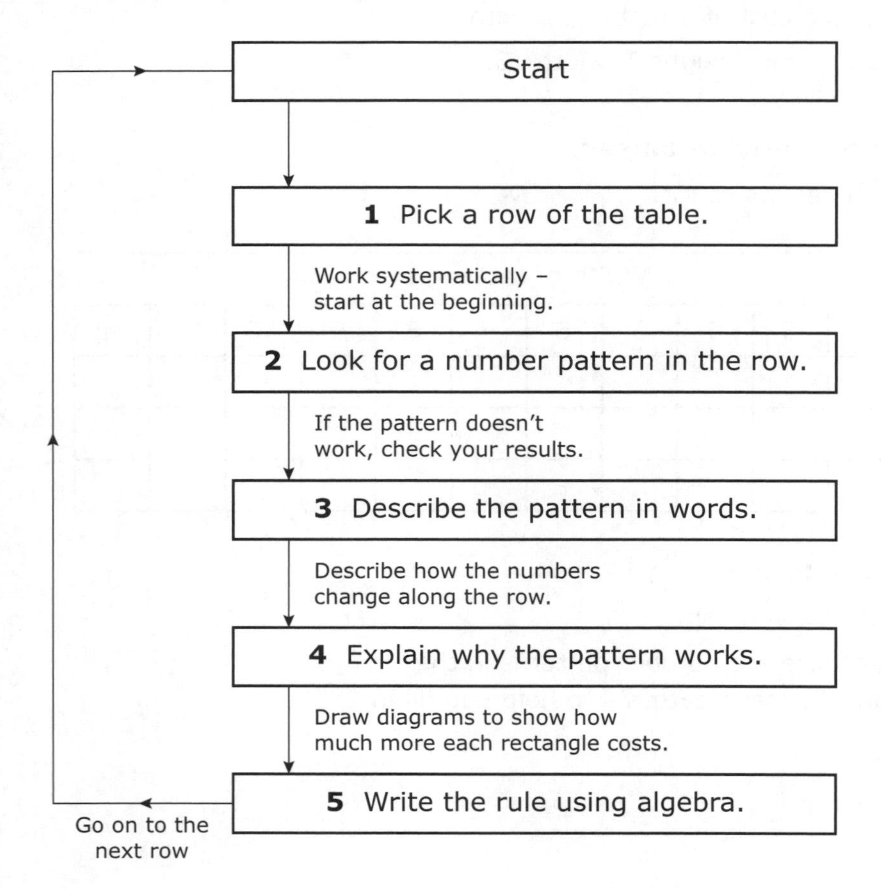

B1.4HW Graphs

In the lesson, you saw how you can use graphs to show your results.

Remember that graphs give you another way to check previous results and predict new ones. If most of your results lie on a straight line when you plot a graph, you would want to check a result that is not on the line!

You should continue plotting graphs to show the results from your table.

Remember:

- Plot the width of the rectangles on the horizontal axis.
- Plot the cost of each rectangle on the vertical axis.
- Choose the scales for each axis so that the graph will include the widest rectangle you want to include, and the most expensive one.
- Your graph will need a title, and will need to be carefully labelled.
- Your graph gives you another chance to check your results, because the points you plot for each row of the table should make a straight line.

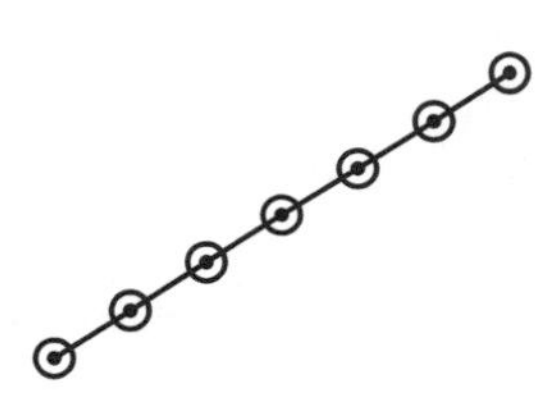

Your lines should look like this.

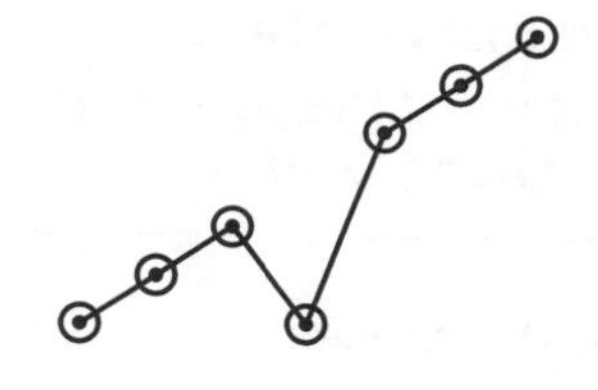

If they look like this, check your results!

B1.5HW Generalising and extending

Continue with the extension to your investigation that you started in the lesson.

1 Asking different questions
Keep the basic situation the same, but extend the problem by changing the questions that you are investigating.

In the original problem, you found the biggest rectangle you could make for £1.

What if the money available was increased to £2? or £3?

You could investigate these questions, and put the results in a table.

Money available	£1	£2	£3	£4	£5
Area of biggest rectangle					

2 Changing the starting information
Change some of the details of the investigation, and see what effect the changes have.

You can change the prices of the pieces.

To find anything significantly different you need to make the changes very clear – like making all the pieces the same price.

3 Major extensions to the problem
Use the original problem as a starting point, and then design a new investigation of your own.

You could extend to investigate 3-D problems.

This kind of extension gives a completely new and more complicated investigation to carry out – make sure you have time to do it!

B1.6HW Reports and conclusions

Continue writing the report of your investigation.

Your finished investigation will probably include these sections:

1 **Title page**

2 **Contents page**

3 **Introduction**
Explain briefly what the investigation is about.

4 **Results**
Explain how you found and checked your results.
Present the results that you need, using tables and graphs.
You only need to include a few examples to show what you did.

5 **Analysis**
Describe and explain the patterns in your results.
Use words and algebra to describe the patterns in your results.
It is very important to try to explain why the rules work.

6 **Conclusions**
Look back at the original questions, and give your answers to them clearly.

In this investigation there was one question with a single answer (finding the largest rectangle you can make for £1), and a more open extension that you could answer in lots of ways.

7 **Appendix**
Include any detailed working and rough work that you want to show, but which would make the main part of the report too cluttered.

> It is a good idea to have page numbers on your report, and to label every chart and diagram (for example 'Table 2', 'Diagram 6').
> This makes it much easier to refer to other parts of your report (for example 'See Diagram 3 on page 6').

If you extend the investigation, you may be able to include your extension work in the sections shown.

If the extension was a major one, you may need to write a separate report!

Level 5

This is a series of patterns with grey and black tiles.

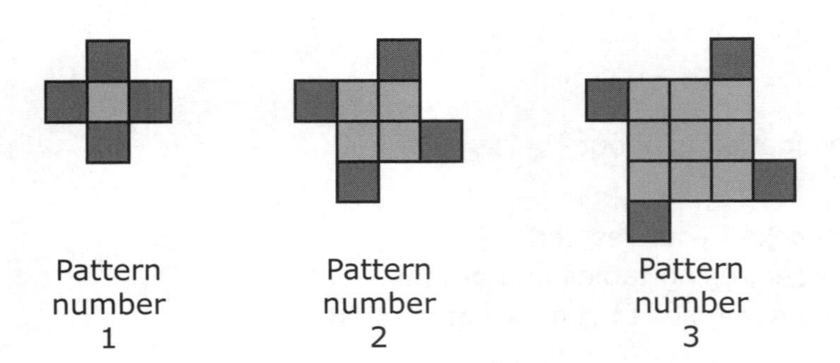

a How many grey tiles and black tiles will there be in pattern number 8? *1 mark*

b How many grey tiles and black tiles will there be in pattern number 16? *1 mark*

c How many grey tiles and black tiles will there be in pattern number P? *1 mark*

d T = total number of grey tiles and black tiles in a pattern
P = pattern number

Use symbols to write down an equation connecting T and P. *1 mark*